Introduction

Pupils' written work
Abacus textbooks are unique in that they provide clear guidance to pupils on how their work should be recorded. Pupils should be encouraged to follow this guidance, which will make marking their work substantially easier, and clearly focused.

Marking pupils' work
Clearly it is important that pupils' work is seen and checked by the teacher regularly, but it is not necessary for all work to be marked by the teacher. Decisions about which work should be teacher-marked, and how it should be marked will be made alongside the need to maximise time available for teaching and guiding pupils through their activities.

A suggested approach within Abacus is to make these decisions Unit by Unit. Decide, for example, for each Unit, which parts you want to mark, and which parts the pupils can mark.

Marking the 'Explores'
The 'Explores' should generally be marked by the teacher. The 'Explores' often require a systematic approach, and the answers give suggestions for these. These approaches can be communicated to the pupils, to help them develop systematic ways of working. Also, the pupils' responses to the 'Explores' may well vary because of the often open-ended nature of the activities.

For many 'Explores' you may want to ask the pupils to work in pairs or groups, possibly leading to a group display of the results of their 'exploration'.

Number Textbook 1

page 3
Hundreds, tens and units

1. 374	2. 258	3. 135	4. 683	5. 491
6. 844	7. 569	8. 958	9. 412	10. 277

● 135   258   277   374   412   491   569   683   844   958

11. 232	233	234	235		12. 145	146	147	148
13. 169	170	171	172		14. 353	354	355	356
15. 420	421	422	423		16. 639	640	641	642
17. 199	200	201	202		18. 898	899	900	901

● 11. 230	229	228	227
12. 143	142	141	140
13. 167	166	165	164
14. 351	350	349	348
15. 418	417	416	415
16. 637	636	635	634
17. 197	196	195	194
18. 896	895	894	893

---

page 4
Hundreds, tens and units

1. 131	2. 143	3. 235	4. 217	5. 111
6. 344	7. 417	8. 370	9. 209	10. 150

● 111   131   143   150   209   217   235   344   370   417

Explore 243 246 247 253 256 257 343 346 347 353 356 357

---

page 5
Hundreds, tens and units

1. 200 + 10 + 9	● two hundred and nineteen
2. 700 + 20 + 4	seven hundred and twenty-four
3. 100 + 50 + 8	one hundred and fifty-eight
4. 300 + 60 + 6	three hundred and sixty-six
5. 900 + 40 + 2	nine hundred and forty-two
6. 400 + 40 + 5	four hundred and forty-five
7. 200 + 20 + 2	two hundred and twenty-two
8. 500 + 70 + 3	five hundred and seventy-three
9. 800 + 30 + 1	eight hundred and thirty-one
10. 700 + 80 + 7	seven hundred and eighty-seven

Number Textbook 1

page 5 cont ...

11. 900 + 6	nine hundred and six			
12. 500 + 60	five hundred and sixty			
13. 700	seven hundred			

14. 524   15. 262   16. 304   17. 810   18. 919
19. 421   20. 639
21. one hundred and eighty-eight
22. seven hundred and seven
23. two hundred and twenty
24. eight hundred and eighty-eight

---

page 6

Hundreds, tens and units

1. 324p	2. 441p	3. 140p	4. 506p	5. 219p
6. 131p	7. 253p	8. 328p	9. 515p	10. 360p

❷ 1. 334p	314p	❷ 2. 451p	431p	❷ 3. 150p	130p
❷ 4. 516p	496p	❷ 5. 229p	209p	❷ 6. 141p	121p
❷ 7. 263p	243p	❷ 8. 338p	318p	❷ 9. 525p	505p
❷ 10. 370p	350p				

All the following have other correct answers.

	£2	£1	50p	20p	10p	5p	2p	1p
11.		1			1		1	1
12.	2			1			1	
13.	1		1				1	
14.	2			1			1	
15.		1				1	1	
16.	1	1		1	1			
17.	3						1	
18.	1		1			1		
19.	1	1				1	1	
20.		1		1				

Number Textbook 1

page 7
Hundreds, tens and units

1. 300 + 80 + 2 = 382
2. 500 + 30 + 8 = 538
3. 100 + 50 = 150
4. 700 + 5 = 705
5. 200 + 70 + 9 = 279
6. 600 + 20 + 1 = 621
7. 900 + 40 + 6 = 946
8. 400 + 4 = 404
9. 700 + 90 = 790
10. forty
11. four hundred
12. seven
13. ten
14. fifty
15. seven hundred
16. sixty
17. three hundred
18. nought

page 8
Hundreds, tens and units

	£1	10p	1p
1.	3	4	1
2.	2	1	2
3.	5	1	3
4.	4	0	5
5.	1	4	8
6.	6	1	5
7.	2	2	0
8.	3	3	3
9.	4	0	9

10. 700 + 30 + 2 = 732
11. 200 + 40 + 5 = 245
12. 800 + 10 + 6 = 816
13. 300 + 80 + 9 = 389
14. 400 + 70 + 1 = 471
15. 900 + 2 = 902
16. 500 + 40 = 540
17. 100 + 60 + 8 = 168
18. 300 + 4 = 304
19. 800 + 80 = 880

Explore
Total 18   smallest 406   largest 964

page 9
Making ten

1. 7
2. 4
3. 8
4. 1
5. 5
6. 3
7. 6
8. 9
9. 10
10. 2

🅔 1. 17   2. 14   3. 18   4. 11   5. 15   6. 13   7. 16   8. 19   9. 20   10. 12

11. 8 + 2 = 10
12. 4 + 6 = 10
13. 7 + 3 = 10
14. 10 + 0 = 10
15. 8 + 2 = 10
16. 5 + 5 = 10
17. 9 + 1 = 10
18. 10 + 0 = 10
19. 6 + 4 = 10
20. 7 + 3 = 10

Number Textbook 1

page 10
Making 20

1. 12 + 8 = 20	2. 16 + 4 = 20	3. 14 + 6 = 20
4. 19 + 1 = 20	5. 11 + 9 = 20	6. 15 + 5 = 20
7. 13 + 7 = 20	8. 18 + 2 = 20	9. 17 + 3 = 20
10. 8 + 12 = 20	11. 14 + 6 = 20	12. 3 + 17 = 20
13. 2 + 18 = 20	14. 15 + 5 = 20	15. 1 + 19 = 20
16. 4 + 16 = 20	17. 13 + 7 = 20	18. 10 + 10 = 20

---

page 11

Making 20

1. 5 + 15 = 20	2. 16 + 4 = 20	3. 13 + 7 = 20
4. 9 + 11 = 20	5. 8 + 12 = 20	6. 11 + 9 = 20
7. 6 + 14 = 20	8. 18 + 2 = 20	9. 14 + 6 = 20
10. 7 + 13 = 20	11. 1 + 19 = 20	12. 10 + 10 = 20
13. 12 + 8 = 20		

❷ Only three sets of three cards can be made without needing two of a specific card.

1, 2, 7
1, 3, 6
1, 4, 5
2, 3, 5

Explore
Pairs

0, 20   1, 19   2, 18   3, 17   4, 16   5, 15   6, 14   7, 13   8, 12   9, 11

You cannot have 10 and 10.

Three cards. Remember there is only one of each card.

1,2,17	2,3,15	3,4,13	4,5,11	5,6,9
1,3,16	2,4,14	3,5,12	4,6,10	5,7,8
1,4,15	2,5,13	3,6,11	4,7,9	
1,5,14	2,6,12	3,7,10		
1,6,13	2,7,11	3,8,9		
1,7,12	2,8,10			
1,8,11				
1,9,10				

Number Textbook 1

page 12
Making 20

1. 14p + 6p = 20p
2. 10p + 10p = 20p
3. 11p + 9p = 20p
4. 18p + 2p = 20p
5. 5p + 15p = 20p
6. 12p + 8p = 20p
7. 9p + 11p = 20p
8. 2p
9. 2p & 1p
10. 10p & 5p & 1p
11. 5p & 2p
12. 5p
13. 5p & 2p & 2p
14. 10p & 2p & 2p

---

page 13
Adding several numbers

1. 12 + 8 + 9 = 29
2. 1 + 8 + 9 = 18
3. 2 + 6 + 8 = 16
4. 4 + 7 + 6 = 17
5. 8 + 7 + 13 = 28
6. 6 + 14 + 6 = 26
7. 18 + 9 + 2 = 29
8. 16 + 7 + 4 = 27
9. 5 + 2 + 15 = 22
10. 10 + 11 + 9 = 30
11. 8 + 2 + 9 = 19
12. 11 + 3 + 7 = 21
13. 14 + 9 + 6 = 29
14. 12 + 9 + 2 = 23
15. 15 + 9 + 5 = 29
16. 15 + 4 + 9 = 28
17. 9 + 4 + 7 = 20
18. 9 + 6 + 3 + 1 = 19
19. 5 + 4 + 10 + 6 = 25
20. 8 + 13 + 2 + 17 = 40

---

page 14
Adding several numbers

Answers will vary.

Explore

10 + 1 + 4	9 + 1 + 5	8 + 1 + 6	7 + 2 + 6	6 + 4 + 5
10 + 2 + 3	9 + 2 + 4	8 + 2 + 5	7 + 3 + 5	
		8 + 3 + 4		

---

page 15
Adding several numbers

1. 10p + 5p + 2p + 1p = 18p
2. 10p + 5p + 2p = 17p
3. 10p + 2p + 2p = 14p
4. 20p + 1p = 21p
5. 10p + 2p = 12p
6. 10p + 5p + 2p + 2p = 19p
7. 20p + 2p + 2p = 24p
8. 10p + 5p = 15p
9. 20p + 5p + 2p = 27p
10. 10p + 2p + 1p = 13p
11. 10p + 5p + 1p = 16p
12. 20p + 5p + 1p = 26p

🖉 Answers will vary.
13. 9 + 3 = 12
14. 12 + 8 − 2 = 18
15. 18 + 6 − 4 = 20
16. 20 + 7 − 3 = 24
17. 24 + 5 − 6 = 23
18. 23 + 7 − 10 = 20

Number Textbook 1

page 16

1. $6 + 4 = 10$
2. $19 + 1 = 20$
3. $28 + 2 = 30$
4. $33 + 7 = 40$
5. $42 + 8 = 50$
6. $56 + 4 = 60$
7. $64 + 6 = 70$
8. $71 + 9 = 80$
9. $87 + 3 = 90$
10. $95 + 5 = 100$
11. $37 + 3 = 40$
12. $45 + 5 = 50$
13. $23 + 7 = 30$
14. $92 + 8 = 100$
15. $18 + 2 = 20$
16. $76 + 4 = 80$
17. $81 + 9 = 90$

page 17

1. $12 + 6 = 18$
2. $24 + 6 = 30$
3. $32 + 7 = 39$
4. $15 + 4 = 19$
5. $22 + 4 = 26$
6. $3 + 36 = 39$
7. $2 + 46 = 48$
8. $25 + 3 = 28$
9. $33 + 5 = 38$
10. $3 + 26 = 29$

e 1. $18 + 2 = 20$
2. $30 + 0 = 30$
3. $39 + 1 = 40$
4. $19 + 1 = 20$
5. $26 + 4 = 30$
6. $39 + 1 = 40$
7. $48 + 2 = 50$
8. $28 + 2 = 30$
9. $38 + 2 = 40$
10. $29 + 1 = 30$

11. $15 + 4 = 19$
12. $17 + 2 = 19$
13. $20 + 4 = 24$
14. $26 + 2 = 28$
15. $48 + 1 = 49$
16. $21 + 3 = 24$
17. $32 + 8 = 40$
18. $51 + 8 = 59$
19. $42 + 4 = 46$
20. $57 + 2 = 59$

page 18

1. $32 + 5 = 37$
2. $43 + 5 = 48$
3. $51 + 5 = 56$
4. $64 + 5 = 69$
5. $21 + 5 = 26$
6. $33 + 5 = 38$
7. $72 + 5 = 77$
8. $94 + 5 = 99$
9. $83 + 5 = 88$
10. $44 + 5 = 49$

e 1. $37 + 5 = 42$
2. $48 + 5 = 53$
3. $56 + 5 = 61$
4. $69 + 5 = 74$
5. $26 + 5 = 31$
6. $38 + 5 = 43$
7. $77 + 5 = 82$
8. $99 + 5 = 104$
9. $88 + 5 = 93$
10. $49 + 5 = 54$

Explore
Answers will vary.

page 19

1. $20p - 4p = 16p$
2. $20p - 3p = 17p$
3. $20p - 9p = 11p$
4. $20p - 6p = 14p$
5. $20p - 8p = 12p$
6. $20p - 1p = 19p$
7. $20p - 2p = 18p$
8. $20p - 11p = 9p$
9. $20p - 5p = 15p$
10. $20p - 7p = 13p$
11. $20p - 14p = 6p$
12. $20p - 13p = 7p$

Number Textbook 1

page 19 cont ...

- a $8 - 5 = 3$
- b $10 - 5 = 5$
- c $16 - 5 = 11$
- d $17 - 5 = 12$
- e $18 - 5 = 13$
- f $20 - 5 = 15$
- g $25 - 5 = 20$
- h $27 - 5 = 22$
- i $29 - 5 = 24$
- j $30 - 5 = 25$

---

page 20

Subtracting

1. $24p - 4p = 20p$
2. $26p - 6p = 20p$
3. $29p - 9p = 20p$
4. $21p - 1p = 20p$
5. $27p - 7p = 20p$
6. $23p - 3p = 20p$
7. $28p - 8p = 20p$
8. $22p - 2p = 20p$
9. $25p - 5p = 20p$
10. $30p - 10p = 20p$
11. $25p + 5p = 30p$ $\qquad$ $25p - 5p = 20p$
12. $34p + 6p = 40p$ $\qquad$ $34p - 4p = 30p$
13. $42p + 8p = 50p$ $\qquad$ $42p - 2p = 40p$
14. $57p + 3p = 60p$ $\qquad$ $57p - 7p = 50p$
15. $63p + 7p = 70p$ $\qquad$ $63p - 3p = 60p$

☻ All the answers are multiples of 10.

---

page 21

1. $27p - 3p = 24p$
2. $26p - 3p = 23p$
3. $25p - 5p = 20p$
4. $20p - 4p = 16p$
5. $47p - 5p = 42p$
6. $19p - 4p = 15p$
7. $20p - 8p = 12p$
8. $35p - 4p = 31p$
9. $25p - 4p = 21p$
10. $37p - 5p = 32p$

11. Tom 11	12. Tom 26	13. Tom 16	14. Tom 24	
15. Gita 12	16. Gita 22	17. Gita 18	18. Gita 30	Gita wins

---

page 22

Subtracting

1. $18 - 3 - 2 = 13$
2. $26 - 1 - 4 = 21$
3. $19 - 5 - 2 = 12$
4. $36 - 2 - 1 = 33$
5. $48 - 5 - 2 = 41$
6. $26 - 4 - 1 = 21$
7. $39 - 7 - 1 = 31$
8. $29 - 5 - 4 = 20$
9. $48 - 6 - 1 = 41$
10. $57 - 3 - 4 = 50$
11. $36 - 2 - 2 = 32$
12. $28 - 2 - 5 = 21$
13. $28p - 7p = 21p$
14. $49p + 49p - 8p = 90p$
15. $15p - 4p = 11p$
16. $10p + 5p = 15p$ therefore Jason has 10p & Mike has 5p

Number Textbook 1

page 23
Counting in ones

1.	489	490	491		1.	484	485	486
2.	668	669	670		2.	663	664	665
3.	291	292	293		3.	286	287	288
4.	110	111	112		4.	105	106	107
5.	986	987	988		5.	981	982	983
6.	848	849	850		6.	843	844	845
7.	545	546	547		7.	540	541	542
8.	166	167	168		8.	161	162	163
9.	499	500	501		9.	494	495	496
10.	330	331	332		10.	325	326	327
11.	719	720	721		11.	714	715	716
12.	198	199	200		12.	193	194	195

57 spots and 22 spots

page 24
Counting back in ones

1.	397	396	395	394	393	392	
2.	252	251	250	249	248	247	
3.	812	811	810	809	808	807	806
4.	402	401	400	399	398		
5.	105	104	103	102	101		
6.	666	665	664	663	662		

Answers will vary.

Pink monster	304	301	402
Blue monster	205	202	303
Yellow monster	104	101	202

page 25
Counting in tens

1.	13	23	33	43	53	63	73	83	93p
2.	7	17	27	37	47	57	67	77	87p
3.	11	21	31	41	51	61	71	81	91p
4.	19	29	39	49	59	69	79	89	99p
5.	2	12	22	32	42	52	62	72	82p

Amos 113p   Jordan 107p   Hannah 111p   Alex 119p   Kim 102p

Number Textbook 1

page 25 cont ...

6.	84	74	64	54	44	34	24	14p
7.	89	79	69	59	49	39	29	19p
8.	76	66	56	46	36	26	16	6p
9.	98	88	78	68	58	48	38	28p
10.	95	85	75	65	55	45	35	25p

page 26
Counting in tens and ones

1.	100	110	120	130	140	150	160	170	180	190	200	210
2.	508	498	488	478	468	458	448	438	428	418	408	
3.	616	626	636	646	656	666	676	686	696	706	716	
4.	333	323	313	303	293	283	273	263	253	243		
5.	89	99	109	119	129	139	149	159	169	179	189	199
6.	111	121	131	141	151	161	171	181	191	201	211	221
7.	686	687	688	689	690	691	692	693	694			
8.	195	194	193	192	191	190	189	188	187			
9.	399	400	401	402	403	404	405	406	407			
10.	809	808	807	806	805	804	803	802	801			
11.	989	990	991	992	993	994	995	996	997			

page 27
Ten less, ten more

1.	317	327	337		2.	635	645	655
3.	777	787	797		4.	828	838	848
5.	406	416	426		6.	976	986	996
7.	202	212	222		8.	383	393	403
9.	495	505	515		10.	890	900	910

❷
1.	227	327	427		2.	545	645	745
3.	687	787	887		4.	738	838	938
5.	316	416	516		6.	886	986	1086
7.	112	212	312		8.	293	393	493
9.	405	505	605		10.	800	900	1000

| 11. | 300 | 600 | | 12. | 350 | 550 |
| 13. | 570 | 770 | | 14. | 710 | 910 |

Number Textbook 1

page 28
Counting in hundreds

I. 400	500	600km
3. 200	300	400km

2. 800	900	1000km
4. 600	700	800km

Explore   The digit total increases by I.

---

page 29

100 less, 100 more

I. 227	327	427		**2.** 455	555	655
3. 118	218	318		**4.** 799	899	999
5. 505	605	705		**6.** 210	310	410
7. 847	947	1047		**8.** 467	567	667
9. 309	409	509		**10.** 41	141	241

@I. 326	327	328		**2.** 554	555	556
3. 217	218	219		**4.** 898	899	900
5. 604	605	606		**6.** 309	310	311
7. 946	947	948		**8.** 566	567	568
9. 408	409	410		**10.** 140	141	142

Jan 122   Feb 222   Mar 212   Apr 242   May 142

---

page 30

I. $3 \times 5 = 15$	**2.** $2 \times 4 = 8$	**3.** $4 \times 3 = 12$	**4.** $5 \times 6 = 30$
5. $4 \times 4 = 16$	**6.** $5 \times 2 = 10$	**7.** $6 \times 3 = 18$	**8.** $3 \times 7 = 21$
9. $3 \times 3 = 9$	**10.** $3 \times 4 = 12$	**II.** $2 \times 5 = 10$	**12.** $5 \times 3 = 15$
13. $2 \times 10 = 20$	**14.** $4 \times 4 = 16$	**15.** $5 \times 2 = 10$	**16.** $2 \times 7 = 14$
17. $3 \times 6 = 18$	**18.** $4 \times 5 = 20$		

@ 10. 24   **II.** 20   **12.** 30   **13.** 40   **14.** 32   **15.** 20   **16.** 28   **17.** 36   **18.** 40

---

page 31

I. $8 \div 4 = 2$	**2.** $12 \div 3 = 4$	**3.** $6 \div 2$ or $6 \div 3$	
4. $15 \div 5 = 3$	**5.** $10 \div 2 = 5$	**6.** $16 \div 4 = 4$	**7.** $12 \div 6 = 2$
8. $8 \div 2 = 4$	**9.** $9 \div 3 = 3$	**10.** $12 \div 4 = 3$	**II.** $10 \div 5 = 2$
12. $15 \div 3 = 5$	**13.** $12 \div 2 = 6$	**14.** $20 \div 4 = 5$	**15.** $20 \div 5 = 4$
16. $10 \div 2 = 5$			

Number Textbook 1

page 32

I. $4 \times 2 = 8$	$8 \div 2 = 4$	**2.** $3 \times 3 = 9$	$9 \div 3 = 3$
3. $4 \times 4 = 16$	$16 \div 4 = 4$	**4.** $2 \times 5 = 10$	$10 \div 5 = 2$
5. $5 \times 3 = 15$	$15 \div 3 = 5$	**6.** $3 \times 2 = 6$	$6 \div 2 = 3$

7. $15 \div 5 = 3$ **8.** $12 \div 3 = 4$ **9.** $18 \div 6 = 3$ **10.** $4 \times 5 = 20$

page 33

I. $40 \div 5 = 8$	**2.** $55 \div 5 = 11$	**3.** $70 \div 5 = 14$	**4.** $35 \div 5 = 7$
5. $12 \div 2 = 6$	**6.** $22 \div 2 = 11$	**7.** $16 \div 2 = 8$	**8.** $18 \div 2 = 9$
9. $3 \times 4 = 12$	**10.** $4 \times 4 = 16$	**II.** $5 \times 3 = 15$	**12.** $2 \times 7 = 14$
13. $6 \times 3 = 18$	**14.** $3 \times 3 = 9$	**15.** $4 \times 6 = 24$	**16.** $8 \times 2 = 16$
17. $2 \times 4 = 8$	**18.** $5 \times 4 = 20$		

🄮 Answers will vary.

Explore

48×1 $= 24 \times 2$ $= 16 \times 3$ $= 12 \times 4$ $= 8 \times 6$

$48 \div 48 = 1$	$48 \div 1 = 48$
$48 \div 24 = 2$	$48 \div 2 = 24$
$48 \div 16 = 3$	$48 \div 3 = 16$
$48 \div 12 = 4$	$48 \div 4 = 12$
$48 \div 8 = 6$	$48 \div 6 = 8$

page 34

I. $5 \times 2 = 10$	**2.** $2 \times 2 = 4$	**3.** $3 \times 2 = 6$	**4.** $6 \times 2 = 12$
5. $4 \times 2 = 8$	**6.** $7 \times 2 = 14$	**7.** $8 \times 2 = 16$	**8.** $10 \times 2 = 20$
9. $9 \times 2 = 18$			

🄮 **I.** $10 \times 2 = 20$	**2.** $4 \times 2 = 8$	**3.** $6 \times 2 = 12$	**4.** $12 \times 2 = 24$
5. $8 \times 2 = 16$	**6.** $14 \times 2 = 28$	**7.** $16 \times 2 = 32$	**8.** $20 \times 2 = 40$
9. $18 \times 2 = 36$			

10. $1 \times 2 = 2$	**II.** $6 \times 2 = 12$	**12.** $4 \times 2 = 8$	**13.** $5 \times 2 = 10$
14. $9 \times 2 = 18$	**15.** $7 \times 2 = 14$	**16.** $10 \times 2 = 20$	**17.** $3 \times 2 = 6$
18. $2 \times 2 = 4$	**19.** $8 \times 2 = 16$		

🄮 $1 \times 2 = 2$ $2 \times 2 = 4$ $3 \times 2 = 6$ $4 \times 2 = 8$ $5 \times 2 = 10$ $6 \times 2 = 12$
$7 \times 2 = 14$ $8 \times 2 = 16$ $9 \times 2 = 18$

Number Textbook 1

page 35

1. $4 \times 2p = 8p$
2. $3 \times 2p = 6p$
3. $7 \times 2p = 14p$
4. $2 \times 2p = 4p$
5. $9 \times 2p = 18p$
6. $5 \times 2p = 10p$
7. $10 \times 2p = 20p$
8. $8 \times 2p = 16p$
9. $6 \times 2p = 12p$

@ 1. 16p 2. 12p 3. 28p 4. 8p 5. 36p 6. 20p 7. 40p 8. 32p 9. 24p

10. $4 \times 2p = 8p$
11. $7 \times 2p = 14p$
12. $5 \times 2p = 10p$
13. $3 \times 2p = 6p$
14. $8 \times 2p = 16p$
15. $2 \times 2p = 4p$
16. $10 \times 2p = 20p$
17. $9 \times 2p = 18p$
18. $6 \times 2p = 12p$

---

page 36

1. $8 \div 2 = 4$
2. $6 \div 2 = 3$
3. $14 \div 2 = 7$
4. $12 \div 2 = 6$
5. $4 \div 2 = 2$
6. $10 \div 2 = 5$
7. $16 \div 2 = 8$
8. $20 \div 2 = 10$
9. $18 \div 2 = 9$
10. $40 \div 2 = 20$

Explore
The numbers coloured should all be in the x2 table.
Patterns described will vary between solid columns of coloured squares (even grids) and, alternate coloured not coloured for rows and columns (odd grids).

---

page 37

Doubling

1. double 6 = 12
2. double 5 = 10
3. double 7 = 14
4. double 3 = 6
5. double 10 = 20
6. double 8 = 16
7. double 2 = 4
8. double 4 = 8
9. double 9 = 18

@ 1. 24 2. 20 3. 28 4. 12 5. 40 6. 32 7. 8 8. 16 9. 36

10. double 4 = 8
11. double 3 = 6
12. double 7 = 14
13. double 14 = 28
14. double 12 = 24
15. double 15 = 30
16. double 6 = 12

---

page 38

Halving and doubling

1. double 7p = 14p
2. double 15p = 30p
3. double 30p = 60p
4. double 40p = 80p
5. double 35p = 70p
6. double 60p = 120p
7. double 45p = 90p
8. double 25p = 50p

Number Textbook 1

page 38 cont ...

doubling	6	3	5	11	15	12	20	30	25	45
	12	6	10	22	30	24	40	60	50	90

halving	4	14	10	16	20	40	30	100	50	70
	2	7	5	8	10	20	15	50	25	35

page 39
Halving and doubling

1. half of 14p = 7p
2. half of 22p = 11p
3. half of 16p = 8p
4. half of 10p = 5p
5. half of 30p = 15p
6. half of 18p = 9p
7. half of 50p = 25p
8. half of £1.40 = 70p
9. half of £2.50 = £1.25
10. double 13p = 26p
11. double 14p = 28p
12. double 25p = 50p
13. double 55p = 110p
14. double 16p = 32p
15. double £1.25 = £2.50
16. double 35p = 70p
17. double 18p = 36p
18. double 65p = 130p

page 40
Fractions

1. 1 square coloured
2. 2 squares coloured
3. 3 squares coloured
4. 2 squares coloured
5. 4 squares coloured
6. 3 squares coloured
7. 1 square coloured

8. $\frac{1}{8}$
9. $\frac{1}{4}$
10. $\frac{1}{4}$
11. $\frac{1}{4}$
12. $\frac{1}{2}$
13. $\frac{1}{4}$
14. $\frac{1}{2}$
15. $\frac{1}{4}$
16. $\frac{1}{8}$
17. $\frac{1}{2}$
18. $\frac{1}{4}$
19. $\frac{1}{8}$

page 41
Fractions

1. $\frac{1}{4}$ of 8 = 2
2. $\frac{1}{2}$ of 6 = 3
3. $\frac{1}{4}$ of 12 = 3
4. $\frac{1}{4}$ of 16 = 4
5. $\frac{1}{2}$ of 20 = 10
6. $\frac{1}{2}$ of 16 = 8
7. $\frac{1}{4}$ of 20 = 5
8. $\frac{1}{3}$ of 6 = 2
9. $\frac{1}{3}$ of 9 = 3
10. $\frac{1}{3}$ of 12 = 4
11. $\frac{1}{3}$ of 15 = 5
12. $\frac{1}{3}$ of 18 = 6

Number Textbook 1

page 42
Fractions

1. $\frac{1}{4}$ of 20cm = 5cm   2. $\frac{1}{4}$ of 24cm = 6cm   3. $\frac{1}{4}$ of 4cm = 1cm

4. $\frac{1}{4}$ of 8cm = 2cm   5. $\frac{1}{4}$ of 32cm = 8cm   6. $\frac{1}{4}$ of 40cm = 10cm

7. $\frac{1}{4}$ of 12cm = 3cm   8. $\frac{1}{3}$ of 6 = 2   9. $\frac{1}{4}$ of 8 = 2

10. $\frac{1}{2}$ of 14 = 7   11. $\frac{1}{3}$ of 21 = 7   12. $\frac{1}{4}$ of 16 = 4

13. $\frac{1}{2}$ of 28 = 14   14. $\frac{1}{3}$ of 36 = 12

Explore

red numbers	3	6	9	12	15	18	21	24	27	30
	33	36								
blue numbers	2	4	6	8	10	12	14	16	18	20
	22	24	26	28	30	32	34	36		

page 43
Fractions

1. $\frac{1}{2}$ of 6 = 3   2. $\frac{1}{3}$ of 9 = 3   3. $\frac{1}{3}$ of 15 = 5

4. $\frac{1}{2}$ of 14 = 7   5. $\frac{1}{3}$ of 12 = 4   6. $\frac{1}{2}$ of 20 = 10

7. $\frac{1}{4}$ of 12 = 3   8. $\frac{1}{4}$ of 16 = 4   9. $\frac{1}{2}$ of 10 = 5

10. $\frac{1}{4}$ of 8 = 2   11. $\frac{1}{2}$ of 12 = 6

12. $\frac{1}{4}$ of 16 = 4 therefore 16 − 4   12 not red   13. $\frac{1}{3}$ of 24 = 8

14. $\frac{7}{14} = \frac{1}{2}$ 15. $\frac{1}{4}$ of 400 = 100

16. $\frac{1}{4}$ of 24 = 6   therefore 24 − 6 = 18 grown-ups

page 44
Adding multiples of ten

1. 30 + 40 = 70   2. 20 + 50 = 70   3. 40 + 40 = 80
4. 60 + 40 = 100   5. 60 + 20 = 80   6. 40 + 50 = 90
7. 60 + 60 = 120   8. 50 + 30 = 80   9. 20 + 70 = 90

10. 30cm + 40cm = 70cm   yes   11. 50cm + 50cm = 100cm   yes
12. 50cm + 60cm = 110cm   no   13. 40cm + 40cm = 80cm   yes
14. 40cm + 50cm = 90cm   yes

Number Textbook 1

page 45
Subtracting multiples of ten

1. 60ml − 30ml = 30ml
2. 70ml − 30ml = 40ml
3. 80ml − 30ml = 50ml
4. 40ml − 30ml = 10ml
5. 90ml − 30ml = 60ml
6. 30ml − 30ml = 0ml

7. 90 − 20 = 70
8. 60 − 30 = 30
9. 70 − 50 = 20
10. 100 − 40 = 60
11. 140 − 30 = 110
12. 270 − 50 = 220
13. 390 − 70 = 320
14. 120 − 30 = 90
15. 150 − 60 = 90

@ 7. 50 8. 10 9. 0 10. 40 11. 90 12. 200 13. 300 14. 70 15. 70

page 46

Adding multiples of five

1. 35 + 20 = 55
2. 45 + 15 = 60
3. 75 + 20 = 95
4. 65 + 25 = 90
5. 35 + 35 = 70
6. 45 + 35 = 80
7. 65 + 40 = 105
8. 85 + 5 = 90
9. 75 + 25 = 100
10. 55 + 35 = 90

11. 35p + 25p = 60p   35p + 65p = 100p   35p + 75p = 110p
    35p + 15p = 50p   35p + 45p = 80p   25p + 45p = 70p
    25p + 65p = 90p   25p + 75p = 100p   25p + 15p = 40p
    65p + 75p = 140p 65p + 15p = 80p   65p + 45p = 110p
    75p + 15p = 90p   75p + 45p = 120p   15p + 45p = 60p

@ pairs to buy with £1

35 & 25   35 & 65   35 & 15   35 & 45   25 & 45   25 & 65
25 & 75   25 & 15   65 & 15   75 & 15   15 & 45

page 47

Subtracting multiples of five

1. 65cm − 30cm = 35cm
2. 70cm − 30cm = 40cm
3. 45cm − 30cm = 15cm
4. 75cm − 30cm = 45cm
5. 55cm − 30cm = 25cm
6. 100cm − 30cm = 70cm
7. 95cm − 30cm = 65cm
8. 35cm − 30cm = 5cm
9. 85cm − 30cm = 55cm
10. 105cm − 30cm = 75cm

Number Textbook 1

page 47 cont ...

II. 65p – 25p = 40p   12. 35p – 25p = 10p   13. 75p – 25p = 50p
14. 60p – 25p = 35p   15. 55p – 25p = 30p   16. 85p – 25p = 60p
17. 40p – 25p = 15p   18. 50p – 25p = 25p

@ II. 40p – 10p = 30p   12. 10p – 10p = 0p   13. 50p – 10p = 40p
   14. 35p – 10p = 25p   15. 30p – 10p = 20p   16. 60p – 10p = 50p
   17. 15p – 10p = 5p   18. 25p – 10p = 15p

---

page 48
Addition pairs to 100

a and e   b & i   c & j   d & f   g & h

1. 10 + 90 = 100   2. 60 + 40 = 100   3. 50 + 50 = 100
4. 15 + 85 = 100   5. 75 + 25 = 100   6. 45 + 55 = 100
7. 75 + 25 = 100   8. 30 + 70 = 100   9. 65 + 35 = 100
10. 95 + 5 = 100   II. 10 + 90 = 100   12. 85 + 15 = 100

@ Answers will vary.

---

page 49
One hundred

1. 100cm – 60cm = 40cm   2. 100cm – 30cm = 70cm
3. 100cm – 80cm = 20cm   4. 100cm – 75cm = 25cm
5. 100cm – 45cm = 55cm   6. 100cm – 10cm = 90cm
7. 100cm – 85cm = 15cm   8. 100cm – 15cm = 85cm
9. 100cm – 95cm = 5cm

10. 65cm   II. 25cm   12. 10cm   13. 45cm   14. 40cm   15. 15cm   16. 60cm
17. 95cm   18. 2cm

@ 10. 115cm   II. 75cm   12. 60cm   13. 95cm   14. 90cm   15. 65cm   16. 110cm
   17. 145cm   18. 52cm

Number Textbook 1

page 50

1. 55p + 45p = £1
2. 35p + 65p = £1
3. 25p + 75p = £1
4. 95p + 5p = £1
5. 85p + 15p = £1
6. 70p + 30p = £1
7. 15p + 85p = £1
8. 45p + 55p = £1
9. 65p + 35p = £1
10. 75p + 25p = £1
11. 20p + 80p = £1
12. 5p + 95p = £1

@ 1. 145p 2. 165p 3. 175p 4. 105p 5. 115p 6. 130p 7. 185p
8. 155p 9. 135p 10. 125p 11. 180p 12. 195p

Explore

98p (9 x 10p + 4 x 2p) + 2p (1 x 2p)
96p (9 x 10p + 3 x 2p) + 4p (2 x 2p)
94p (9 x 10p + 2 x 2p) + 6p (3 x 2p)
92p (9 x 10p + 1 x 2p) + 8p (4 x 2p)
90p (9 x 10p) + 10p (5 x 2p)
90p (8 x 10p + 5 x 2p) + 10p (1 x 10p)
88p (8 x 10p + 4 x 2p) + 12p (1 x 10p + 1 x 2p)
86p (8 x 10p + 3 x 2p) + 14p (1 x 10p + 2 x 2p)
... and so on. Encourage children to use this systematic approach.

---

page 51
Hundreds, tens and units

1. a = 314   315   b = 325   326   c = 333   334   d = 341   342
   e = 348   349
2. f = 654   655   g = 662   663   h = 678   679   i = 685   686
   j = 696   697

@ a. 324 b. 335 c. 343 d. 351 e. 358 f. 664 g. 672 h. 688
   i. 695 j. 706

3. 255   246   156   4. 477   468   378
5. 346   337   247   6. 208   199   109
7. 195   186   96   8. 304   295   205
9. 110   101   11

Number Textbook 1

page 52
Hundreds, tens and units

I.	210	211	212	**2.**	476	477	478
3.	907	908	909	**4.**	316	317	318
5.	278	279	280	**6.**	100	101	102
7.	899	900	901	**8.**	400	401	402
9.	330	331	332	**10.**	222	223	224
II.	831	832	833	**12.**	510	511	512
13.	601	602	603				

Explore

109	119	129	139	149	159	169	179	189	190	191
192	193	194	195	196	197	198	199			
209	219	229	239	249	259	269	279	289	290	291
292	293	294	295	296	297	298	299			

page 53

Smallest to largest

I.	210	532	987	**2.**	402	673	999
3.	241	336	372	**4.**	340	560	600
5.	245	255	325	**6.**	289	299	399
7.	772	778	787	**8.**	411	441	444
9.	178	781	871	**10.**	101	110	111

II. largest 599	smallest 301	**12.** largest 444	smallest 111
13. largest 929	smallest 819	**14.** largest 606	smallest 206
15. largest 538	smallest 531	**16.** largest 775	smallest 720
17. largest 630	smallest 306	**18.** largest 501	smallest 105

@ Answers will vary.

page 54

Hundreds, tens and units

Answers from the ranges:

I.	425, ... 435	**2.**	401, ... 409	**3.**	771, ... 789
4.	556, ... 565	**5.**	329, ... 341	**6.**	205, ... 213
7.	788, ... 794	**8.**	697, ... 702	**9.**	509, ... 517
10.	257 and 258				

Number Textbook 1
page 54 cont ...
Explore

£6 (6 x £1)
£5.10 (5 x £1 + 1 x 10p)
£5.01 (5 x £1 + 1 x 1p)
£4.20p (4 x £1 + 2 x 10p)
£4.11p (4 x £1 + 1 x 10p + 1 x 1p)
£4.02 (4 x £1 + 2x 1p)
£3.30 (3 x £1 + 3x 10p)
£3.21 (3 x £1 + 2 x 10p + 1 x 1p)
£3.12 (3 x £1 + 1 x 10p + 2 x 1p)
£3.03 (3 x £1 + 3 x 1p)
£2.40 (2 x £1 + 4 x 10p)

£2.31 (2 x £1 + 3 x 10p + 1 x 1p)
£2.22 (2 x £1 + 2 x 10p + 2 x 1p)
£2.13 (2 x £1 + 1 x 10p + 3 x 1p)
£2.04 (2 x £1 + 4 x 1p)
£1.50 (1 x £1 + 5 x 10p)
£1.41 (1 x £1 + 4 x 10p + 1 x 1p)
£1.32 (1 x £1 + 3 x 10p + 2 x 1p)
£1.23 (1 x £1 + 2 x 10p + 3 x 1p)
£1.14 (1 x £1 + 1 x 10p + 4 x 1p)
£1.05 (1 x £1 + 5 x 1p)

---

page 55
Addition **N16**
Adding to the next ten

1. $5 + 5 = 10$
2. $4 + 6 = 10$
3. $1 + 9 = 10$
4. $2 + 8 = 10$
5. $7 + 3 = 10$
6. $0 + 10 = 10$
7. $8 + 2 = 10$
8. $16 + 4 = 20$
9. $23 + 7 = 30$
10. $28 + 2 = 30$
11. $37 + 3 = 40$
12. $42 + 8 = 50$
13. $46 + 4 = 50$
14. $55 + 5 = 60$
15. $64 + 6 = 70$
16. $69 + 1 = 70$
17. $74 + 6 = 80$
18. $85 + 5 = 90$
19. $88 + 2 = 90$
20. $92 + 8 = 100$
21. $99 + 1 = 100$

---

page 56
Addition **N16**
Adding

1. $27 + 3 = 30$ $27 + 5 = 32$
2. $37 + 3 = 40$ $37 + 4 = 41$
3. $45 + 5 = 50$ $45 + 7 = 52$
4. $58 + 2 = 60$ $58 + 4 = 62$
5. $63 + 7 = 70$ $63 + 8 = 71$
6. $26 + 4 = 30$ $26 + 6 = 32$
7. $44 + 6 = 50$ $44 + 8 = 52$
8. $15 + 5 = 20$ $15 + 8 = 23$
9. $78 + 2 = 80$ $78 + 3 = 81$
10. $36 + 4 = 40$ $36 + 8 = 44$
11. $57 + 3 = 60$ $57 + 5 = 62$
12. $69 + 1 = 70$ $69 + 4 = 73$
13. $88 + 2 = 90$ $88 + 8 = 96$

14. $16cm + 6cm = 22cm$
15. $18cm + 6cm = 24cm$
16. $26cm + 6cm = 32cm$
17. $25cm + 6cm = 31cm$
18. $35cm + 6cm = 41cm$
19. $17cm + 6cm = 23cm$

@ 14. $22cm \times 2 = 44cm$
15. $24cm \times 2 = 48cm$
16. $32cm \times 2 = 64cm$
17. $31cm \times 2 = 62cm$
18. $41cm \times 2 = 82cm$
19. $23cm \times 2 = 46cm$

Number Textbook 1

page 57
Adding

1. 16p + 5p = 21p	2. 27p + 5p = 32p	3. 36p + 5p = 41p
4. 45p + 5p = 50p	5. 29p + 5p = 34p	6. 38p + 5p = 43p
7. 47p + 5p = 52p	8. 56p + 5p = 61p	9. 66p + 5p = 71p
10. 77p + 5p = 82p		

Explore

87 + 6 = 93	86 + 7 = 93	85 + 7 = 92	78 + 6 = 84
87 + 5 = 92	86 + 5 = 91	85 + 6 = 91	78 + 5 = 83
76 + 8 = 84	75 + 8 = 83	58 + 7 = 65	57 + 8 = 65
76 + 5 = 81	75 + 6 = 81	58 + 6 = 64	57 + 6 = 63
56 + 8 = 64			
56 + 7 = 63			

18 different additions.

9 different answers.

Children should notice that the answers are the same when the same tens digit occurs but the units digits are reversed i.e. 56 + 7 and 57 + 6.

---

page 58
Difference

1. 30p + 4p = 34p	2. 40p + 6p = 46p	3. 50p + 5p = 55p
4. 40p + 9p = 49p	5. 20p + 3p = 23p	6. 10p + 8p = 18p
7. 60p + 6p = 66p	8. 50 + 4 = 54	9. 20 + 8 = 28
10. 40 + 6 = 46	11. 70 + 7 = 77	12. 60 + 5 = 65
13. 80 + 9 = 89		

---

page 59
Difference

1. 87 + 3 = 90	2. 49 + 1 = 50	3. 26 + 4 = 30	4. 55 + 5 = 60
5. 68 + 2 = 70	6. 33 + 7 = 40	7. 14 + 6 = 20	8. 82 + 8 = 90
9. 41 + 9 = 50	10. 32 + 8 = 40	11. 33 − 27 = 6	12. 54 − 47 = 7
13. 34 − 28 = 6	14. 61 − 56 = 5	15. 44 − 37 = 7	16. 62 − 58 = 4

Number Textbook 1

page 60
Difference

I. $32 - 28 = 4$ 2. $43 - 36 = 7$ 3. $52 - 46 = 6$ 4. $24 - 17 = 7$
5. $52 - 44 = 8$ 6. $41 - 36 = 5$ 7. $35 - 29 = 6$ 8. $26 - 15 = 11$
9. $32 - 26 = 6$ 10. $45 - 37 = 8$

II. $23kg - 16kg = 7kg$ I2. $33kg - 28kg = 5kg$ I3. $21kg - 17kg = 4kg$

Explore II and 7

---

page 61
Difference

I. $33 - 27 = 6$ 2. $42 - 38 = 4$ 3. $22 - 18 = 4$ 4. $31 - 26 = 5$
5. $23 - 19 = 4$ 6. $32 - 27 = 5$ 7. $53 - 45 = 8$ 8. $53 - 46 = 7$
9. $32 - 25 = 7$ 10. $44 - 36 = 8$ II. $24 - 18 = 6$ I2. $43 - 39 = 4$
I3. $51 - 49 = 2$

◉ Answers will vary.

I4. $23p - 18p = 5p$ I5. $51p - 45p + 30p = 36p$
I6. $26p + 8p - 29p + 50p = 55p$ I7. $32p - 25p + 40p = 47p$

---

page 62
Subtracting multiples of ten

I. 92   82   72   62   52   42   32
2. 96   86   76   66   56   46   36   26   16   6
3. 98   88   78   68   58   48   38   28   18   8
4. 91   81   71   61   51   41   31   21   11   1

5. $84p - 20p = 64p$ 6. $31p - 20p = 11p$ 7. $45p - 20p = 25p$
8. $£1.00 - 20p = 80p$ 9. $£1.20 - 20p = £1.00$ 10. $£1.30 - 20p = £1.10$

---

page 63
Adding multiples of ten

I. $£1.24 + 30p = £1.54$ 2. $£1.62 + 20p = £1.82$
3. $£1.79 + 20p = £1.99$ 4. $£2.55 + 40p = £2.95$
5. $£3.34 + 40p = £3.74$ 6. $£4.15 + 50p = £4.65$
7. $£5.17 + 20p = £5.37$ 8. $£8.80 + 20p = £9.00$

Number Textbook 1

page 63 cont ...

9. £6.43 + 40p = £6.83
10. £1.40 + 60p = £2
11. £1.50 + 50p = £2
12. £1.30 + 70p = £2
13. £1.70 + 30p = £2
14. £1.20 + 80p = £2
15. £1.10 + 90p = £2
16. £1.75 + 25p = £2
17. £1.45 + 55p = £2

@ 10. £1.10 11. £1.00 12. £1.20 13. 80p 14. £1.30 15. £1.40 16. 75p 17. £1.05

---

page 64

Subtracting multiples of ten

1. £1.55 − 30p = £1.25
2. £1.99 − 30p = £1.69
3. £2.45 − 30p = £2.15
4. £1.63 − 30p = £1.33
5. £3.33 − 30p = £3.03
6. £4.62 − 30p = £4.32
7. £5.50 − 30p = £5.20
8. £2.99 − 30p = £2.69
9. £3.75 − 30p = £3.45
10. £5.81 − 30p = £5.51

@ 1. 75p 2. £1.19 3. £1.65 4. 83p 5. £2.53 6. £3.82 7. £4.70
8. £2.19 9. £2.95 10. £5.01

Explore

The children have to spend £1.70.
These are the possible outcomes, although the order may vary. Encourage them
to use a system.

£1 + 60p + 10p	90p + 70p + 10p	80p + 80p + 10p	70p + 90p + 10p
£1 + 50p + 20p	90p + 60p + 20p	80p + 70p + 20p	70p + 80p + 20p
£1 + 40p + 30p	90p + 50p + 30p	80p + 60p + 30p	70p + 70p + 30p
	90p + 40p + 40p	80p + 50p + 40p	70p + 60p + 40p
			70p + 50p + 50p

---

page 65

Adding and subtracting

1. £1.12 + 10p = £1.22
2. £1.41 + 20p = £1.61
3. £1.38 + 40p = £1.78
4. £1.29 + 20p = £1.49
5. £1.73 − 10p = £1.63
6. £1.92 − 30p = £1.62
7. £3.16 + 20p = £3.36
8. £4.44 − 30p = £4.14
9. £5.12 + 10p = £5.22
10. £4.71 − 30p = £4.41
11. £3.27 + 50p = £3.77
12. £2.37 + 40p = £2.77
13. £1.52 − 30p = £1.22
14. £5.21 + 50p = £5.71   saved 50p

Number Textbook 1

page 66
Adding

1. $17 + 30 = 47$
2. $37 + 20 = 57$
3. $34 + 40 = 74$
4. $48 + 30 = 78$
5. $29 + 40 = 69$
6. $56 + 33 = 89$
7. $62 + 23 = 85$
8. $38 + 31 = 69$
9. $47 + 42 = 89$
10. $46 + 23 = 69$

11. All the following are possible.

$22p + 16p = 38p$ $22p + 70p = 92p$ $22p + 11p = 33p$
$22p + 13p = 35p$ $22p + 24p = 46p$ $22p + 63p = 85p$
$22p + 12p = 34p$ $16p + 70p = 86p$ $16p + 11p = 27p$
$16p + 13p = 29p$ $16p + 24p = 40p$ $16p + 63p = 79p$
$16p + 12p = 28p$ $70p + 11p = 81p$ $70p + 13p = 83p$
$70p + 24p = 94p$ $70p + 63p = 133p$ $70p + 12p = 82p$
$13p + 24p = 37p$ $13p + 63p = 76p$ $13p + 12p = 25p$
$24p + 63p = 87p$ $24p + 12p = 36p$

page 67
Adding

1. $123 + 41 = 164$
2. $135 + 41 = 176$
3. $130 + 54 = 184$
4. $123 + 54 = 177$
5. $135 + 54 = 189$
6. $62 + 123 = 185$
7. $130 + 62 = 192$
8. $62 + 126 = 188$
9. $41 + 126 = 167$
10. $135 + 62 = 197$

⊘ Answers will vary.

Explore

$14kg + 24kg = 38kg$
$27kg + 12kg = 39kg$
$26kg + 13kg = 39kg$
$22kg + 17kg = 39kg$

page 68
Subtracting

1. $£1.45 - 21p = £1.24$
2. $£2.74 - 32p = £2.42$
3. $£2.55 - 42p = £2.13$
4. $£3.52 - 31p = £3.21$
5. $£1.66 - 35p = £1.31$
6. $£2.38 - 23p = £2.15$
7. $£3.41 - 31p = £3.10$
8. $68 - 21 + 33 = 80$
9. $121 + 42 - 51 = 112$
10. $136 + 32 - 53 = 115$

Number Textbook 1

page 69
Adding

1. 32 + 29 = 61 litres
2. 44 + 29 = 73 litres
3. 25 + 29 = 54 litres
4. 29 + 29 = 58 litres
5. 34 + 29 = 63 litres
6. 68 + 29 = 97 litres
7. 77 + 29 = 106 litres
8. 48 + 29 = 77 litres
9. 53 + 29 = 82 litres
10. 36 + 29 = 65 litres

@ 1. 61 + 19 = 80 litres
2. 73 + 19 = 92 litres
3. 54 + 19 = 73 litres
4. 58 + 19 = 77 litres
5. 63 + 19 = 82 litres
6. 97 + 19 = 116 litres
7. 106 + 19 = 125 litres
8. 77 + 19 = 96 litres
9. 82 + 19 = 101 litres
10. 65 + 19 = 84 litres

Explore
The number and its reverse always add to a multiple of 11.

---

page 70
Subtracting

1. 34 – 9 = 25
2. 46 – 9 = 37
3. 53 – 19 = 34
4. 21 – 9 = 12
5. 45 – 19 = 26
6. 58 – 19 = 39
7. 73 – 29 = 44
8. 66 – 29 = 37
9. 47 – 39 = 8
10. 33 – 19 = 14
11. 51 – 39 = 12
12. 78 – 49 = 29
13. 61 – 49 = 12

@ 1. 134
2. 146
3. 143
4. 121
5. 135
6. 148
7. 153
8. 146
9. 117
10. 123
11. 121
12. 138
13. 121

Explore
12 is the closest to 11 you can get.
21 – 9 = 12   31 – 19 = 12   41 – 29 = 12

---

page 71
Adding

1. 124 miles + 19 miles = 143 miles
2. 132 miles + 19 miles = 151 miles
3. 143 miles + 29 miles = 172 miles
4. 232 miles + 19 miles = 251 miles
5. 146 miles + 29 miles = 175 miles
6. 162 miles + 39 miles = 201 miles

Miss Zippy   42 + 29 + 39 = 110 miles
Miss Pootle   22 + 19 + 29 = 70 miles
@ 110 – 70 = 40 miles further

Number Textbook 1

page 72
Adding and subtracting

1. $15 + 19 = 34$
2. $18 + 29 = 47$
3. $22 + 39 = 61$
4. $45 - 19 = 26$
5. $32 + 39 = 71$
6. $78 - 29 = 49$
7. $27 + 39 = 66$
8. $99 - 29 = 70$
9. $16 + 49 = 65$
10. $83 - 39 = 44$

@ Answers will vary.

11. $150 - (2 \times 39) = 150 - 78 = 72ml$
12. $250 - (2 \times 49) = 250 - 98 = 152cm$
13. £3.32 + 29p = £3.61
14. $158m + 49m = 207m$

Number Textbook 2

page 3
Multiples of 10

1. 140 390 550 90 400 710 800 880 330 680 900
590 1000 950 220
@ Multiples of 100   400   800   900   1000

Explore

220 230 240 250 260 270 280 290 300 310 320
330 340 350 360 370 380 390 400 410 420 430
440 450 (24)
@ Multiples of 100   300   400   (2)

---

page 4
Counting in 50s

1. a 550
2. b 750   c 950   d 1150
3. e 950   f 1150
4. g 1050   h 1250   i 1350
5. j 1350   k 1650

6. 564 664 764   7. 447 547 647
8. 679 779 879   9. 645 745 845
10. 594 694 794   11. 788 888 988
12. 399 499 599   13. 442 542 642
14. 756 856 956   15. 807 907 1007
16. 619 719 819   17. 302 402 502
18. 198 298 398

@ 6. 364   7. 247   8. 479   9. 445
10. 394   11. 588   12. 199   13. 242
14. 556   15. 607   16. 419   17. 102
18. not possible

---

page 5

Counting in 50s

1. 850 900 950 1000   2. 700 750 800 850
3. 350 400 450 500   4. 150 200 250 300
5. 400 450 500 550   6. 850 800 750 700
7. 600 550 500 450

Number Textbook 2

page 5 cont ...

8. £2.50 + 50p = £3.00
10. 25cm + 50cm = 75cm

9. 350g + (3 x 50) = 500g
11. 750m − (5 x 50) = 500m

---

page 6
Odd and even

1. 7	odd	**2.** 3	odd	**3.** 8	even	**4.** 6	even
5. 12	even	**6.** 9	odd	**7.** 4	even		

8. 34	36	38	40	42
9. 10	12	14	16	18
10. 28	30	32	34	36
11. 42	44	46	48	50
12. 7	9	11	13	15
13. 25	27	29	31	33
14. 41	43	45	47	49

---

page 7
Odd and even

1. 42 even	**2.** 16 even	**3.** 43 odd	**4.** 21 odd
5. 34 even	**6.** 49 odd	**7.** 33 odd	**8.** 51 odd
9. 11 odd	**10.** 10 even	**11.** 14 even	**12.** 32 even
13. 50 even	**14.** 25 odd	**15.** 17 odd	**16.** 38 even
17. 47 odd			

Explore
4 odd (23, 25, 35, 53)
2 even (32, 52)

Answers will vary but the pattern is that for two odd or two even the results will always be 2 to 4. Obviously if all three are odd or even all the answers will be either odd or even.

---

page 8
Odd and even

2	4	6	8	10	12	14	16	18	20	22
24	26	28	30	32	34	36	38	40	42	44
46	48	50								

Number Textbook 2

page 8 cont ...

	49	47	45	43	41	39	37	35	33	31	29
	27	25	23	21	19	17	15	13	11	9	7
	5	3	1								

1. 46 or 48
2. 10
3. any of   1   3   5   7   9   11   13 years
4. 5

---

page 9
Addition/subtraction **N23**

Pairs to 100

40 and 60	35 and 65	45 and 55	25 and 75
15 and 85	30 and 70	20 and 80	95 and 5

9. $70 + 30 = 100$
10. $75 + 25 = 100$
11. $65 + 35 = 100$
12. $45 + 55 = 100$
13. $100 - 5 - 95$
14. $100 - 85 = 15$

---

page 10
Addition/subtraction **N23**

Pairs to 100

1. $£1.00 - 30p = 70p$
2. $£1.00 - 75p = 25p$
3. $£1.00 - 80p = 20p$
4. $£1.00 - 45p = 55p$
5. $£1.00 - 60p = 40p$
6. $£1.00 - 55p = 45p$
7. $£1.00 - 85p = 15p$
8. $£1.00 - 90p = 10p$
9. $£1.00 - 65p = 35p$
10. $£1.00 - 25p = 75p$

1. £1.70
2. £1.25
3. £1.20
4. £1.55
5. £1.40
6. £1.45
7. £1.15
8. £1.10
9. £1.35
10. £1.75

Explore

5p	20	18	16	14	12	10	8	6	4	2	0
10p		1	2	3	4	5	6	7	8	9	10

Number Textbook 2

page 11
Adding 2-digit numbers

1. $60 + 40 = 100$
2. $59 + 40 = 99$
3. $35 + 65 = 100$
4. $71 + 32 = 103$
5. $75 + 26 = 101$
6. $85 + 14 = 99$
7. $79 + 12 = 91$
8. $49 + 49 = 98$
9. $75 + 35 = 110$
10. $15 + 87 = 102$
11. $29 + 69 = 98$
12. $61 + 42 = 103$
13. $19 + 90 = 109$
14. £1.25 + 65p = £1.90   £2 − £1.90 = 10p
15. 55p + 45p = £1   £2 − £1 = £1
16. £1 − 75p = 25p
17. 95p + 95p = £1.90   £2 − £1.90 = 10p

---

page 12
Doubling

1. $23 \rightarrow 46$
2. $24 \rightarrow 48$
3. $31 \rightarrow 62$
4. $42 \rightarrow 84$
5. $32 \rightarrow 64$
6. $12 \rightarrow 24$
7. $21 \rightarrow 42$
8. $34 \rightarrow 68$
9. $52 \rightarrow 104$
10. $43 \rightarrow 86$

11. 22p → 44p   44p → 88p
12. 12p → 24p   24p → 48p
13. 32p → 64p   64p → 128p
14. 11p → 22p   22p → 44p
15. 21p → 42p   42p → 84p
16. 20p → 40p   40p → 80p
17. 31p → 62p   62p → 124p

---

page 13
Doubling

1. 95p → 190p → £1.90
2. 35p → 70p
3. 45p → 90p
4. 55p → 110p → £1.10
5. 5p → 10p
6. 15p → 30p
7. 75p → 150p → £1.50
8. 25p → 50p
9. 65p → 130p → £1.30
10. 85p → 170p → £1.70

Explore

105 → 210   205 → 410   305 → 610
405 → 810   505 → 1010   605 → 1210
705 → 1410   805 → 1610   905 → 1810
The answers are all multiples of 10.

Number Textbook 2

page 14
Doubling

1.	25 → 50	26 → 52
2.	15 doubled is 30	16 doubled is 32
3.	35 + 35 = 70	36 + 36 = 72
4.	double 12 is 24	double 13 is 26
5.	45 doubled is 90	46 doubled is 92
6.	25 + 25 = 50	25 + 26 = 51
7.	24 doubled is 48	23 doubled is 46
8.	double 50 is 100	double 52 is 104
9.	50 + 50 = 100	49 + 49 = 98
10.	30 + 30 = 60	29 + 30 = 59
11.	double 100 is 200	double 99 is 198
12.	double 43p is 86p	
13.	double 34p is 68p	
14.	double 102p is 204p (or £2.04)	

page 15
Fives

5	10	15	20	25	30	35	40	45	50	55
60	65	70	75	80	85	90	95	100		

1. 3 x 5 = 15   **2.** 2 x 5 = 10   **3.** 4 x 5 = 20   **4.** 6 x 5 = 30
5. 9 x 5 = 45   **6.** 10 x 5 = 50   **7.** 8 x 5 = 40   **8.** 7 x 5 = 35
9. 5 x 5 = 25   **10.** 1 x 5 = 5   **11.** 4 x 5 = 20   **12.** 5 x 5 = 25
13. 9 x 5 = 45   **14.** 8 x 5 = 40   **15.** 3 x 5 = 15   **16.** 6 x 5 = 30
17. 10 x 5 = 50   **18.** 7 x 5 = 35
19. 2 x 5 = 10

🌀 1 x 5 = 5   2 x 5 = 10   3 x 5 = 15   4 x 5 = 20   5 x 5 = 25
6 x 5 = 30   7 x 5 = 35   8 x 5 = 40   9 x 5 = 45   10 x 5 = 50

page 16
Tens

1. 1 x 10 = 10   **2.** 6 x 10 = 60   **3.** 4 x 10 = 40   **4.** 9 x 10 = 90
5. 3 x 10 = 30   **6.** 7 x 10 = 70   **7.** 5 x 10 = 50   **8.** 10 x 10 = 100
9. 8 x 10 = 80   **10.** 2 x 10 = 20

Number Textbook 2

page 16 cont ...

◉ $1 \times 10 = 10$ $2 \times 10 = 20$ $3 \times 10 = 30$ $4 \times 10 = 40$ $5 \times 10 = 50$
   $6 \times 10 = 60$ $7 \times 10 = 70$ $8 \times 10 = 80$ $9 \times 10 = 90$ $10 \times 10 = 100$

11. $6 \times 10p = 60p$ **12.** $7 \times 10p = 70p$ **13.** $5 \times 10p = 50p$ **14.** $8 \times 10p = 80p$
15. $4 \times 10p = 40p$ **16.** $3 \times 10p = 30p$ **17.** $9 \times 10p = 90p$ **18.** $2 \times 10p = 20p$
19. $1 \times 10p = 10p$

◉ **11.** $6 \times 5p = 30p$ **12.** $7 \times 5p = 35p$ **13.** $5 \times 5p = 25p$ **14.** $8 \times 5p = 40p$
   15. $4 \times 5p = 20p$ **16.** $3 \times 5p = 15p$ **17.** $9 \times 5p = 45p$ **18.** $2 \times 5p = 10p$
   19. $1 \times 5p = 5p$

page 17
Fives and tens

  1. $30p \div 5p = 6$ **2.** $15p \div 5p = 3$ **3.** $5p \div 5p = 1$
  4. $40p \div 5p = 8$ **5.** $25p \div 5p = 5$ **6.** $45p \div 5p = 9$
  7. $10p \div 5p = 2$ **8.** $20p \div 5p = 4$ **9.** $50p \div 5p = 10$
 10. $35p \div 5p = 7$ **11.** $45p \div 5p = 9$ days **12.** $20 \div 5 = 4$ boxes

Explore
5p   10p   15p   20p   25p   30p   35p

page 18
Multiplying

  1. $5 \times 2 = 10$ **2.** $5 \times 4 = 20$ **3.** $5 \times 1 = 5$ **4.** $5 \times 3 = 15$
  5. $2 \times 3 = 6$ **6.** $6 \times 2 = 12$ **7.** $6 \times 1 = 6$ **8.** $4 \times 2 = 8$
  9. $5 \times 6 = 30$ **10.** $4 \times 4 = 16$

11. 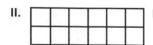   **12.** 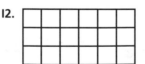   **13.**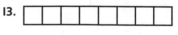

14.   **15.** 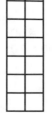   **16.** 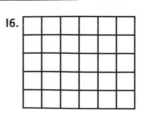   **17.**

18.

Number Textbook 2

page 19
Multiplying

1. $2 \times 3 = 3 \times 2 = 6$
2. $4 \times 3 = 3 \times 4 = 12$
3. $6 \times 2 = 2 \times 6 = 12$
4. $4 \times 4 = 16$
5. $5 \times 3 = 3 \times 5 = 15$
6. $4 \times 5 = 5 \times 4 = 20$
7. $2 \times 6 = 6 \times 2 = 12$
8. $5 \times 5 = 25$
9. $6 \times 4 = 4 \times 6 = 24$
10. $6 \times 3 = 3 \times 6 = 18$

❷ Answers will vary.

Explore
$24 = 1 \times 24 = 2 \times 12 = 3 \times 8 = 4 \times 6$ (4 rectangles)

---

page 20
Multiplying and dividing

1. $8 \div 4 = 2$ $4 \times 2 = 8$
2. $10 \div 5 = 2$ $2 \times 5 = 10$
3. $12 \div 3 = 4$ $3 \times 4 = 12$
4. $6 \div 2 = 3$ $2 \times 3 = 6$
5. $15 \div 3 = 5$ $3 \times 5 = 15$
6. $20 \div 5 - 4$ $5 \times 4 = 20$
7. $12 \div 6 = 2$ $6 \times 2 = 12$
8. $20 \div 10 = 2$ $10 \times 2 = 20$

x	1	2	3	4	5	6
1	1	2	3	4	5	6
2	2	4	6	8	10	12
3	3	6	9	12	15	18
4	4	8	12	16	20	24
5	5	10	15	20	25	30
6	6	12	18	24	30	36

---

page 21
Threes

	3	6	9	12	15	18	21	24	27

1. $4 \times 3 = 12$
2. $2 \times 3 = 6$
3. $6 \times 3 = 18$
4. $5 \times 3 = 15$
5. $3 \times 3 = 9$
6. $8 \times 3 = 24$
7. $7 \times 3 = 21$
8. $9 \times 3 = 27$
9. $10 \times 3 = 30$
10. $1 \times 3 = 3$
11. $5 \times 3 = 15$
12. $9 \times 3 = 27$
13. $6 \times 3 = 18$
14. $3 \times 3 = 9$
15. $2 \times 3 = 6$
16. $10 \times 3 = 30$
17. $8 \times 3 = 24$
18. $4 \times 3 = 12$
19. $7 \times 3 = 21$

❷ $1 \times 3 = 3$ $2 \times 3 = 6$ $3 \times 3 = 9$ $4 \times 3 = 12$ $5 \times 3 = 15$
   $6 \times 3 = 18$ $7 \times 3 = 21$ $8 \times 3 = 24$ $9 \times 3 = 27$ $10 \times 3 = 30$

Number Textbook 2

page 22
Threes

I. $4 \times 3 = 12$ **2.** $6 \times 3 = 18$ **3.** $1 \times 3 = 3$ **4.** $3 \times 3 = 9$
5. $5 \times 3 = 15$ **6.** $7 \times 3 = 21$ **7.** $2 \times 3 = 6$ **8.** $2 \times 3 = 6$
9. $6 \times 3 = 18$ **10.** $5 \times 3 = 15$ **II.** $3 \times 3 = 9$ **12.** $12 \div 3 = 4$
13. $30 \div 3 = 10$ **14.** $3 \div 3 = 1$ **15.** $9 \times 3 = 27$ **16.** $24 \div 3 = 8$
17. $7 \times 3 = 21$

page 23
Threes

I. $10 \times 3 = 30$ **2.** $6 \times 3 = 18$ **3.** $4 \times 3 = 12$ **4.** $9 \times 3 = 27$
5. $2 \times 3 = 6$ **6.** $1 \times 3 = 3$ **7.** $8 \times 3 = 24$ **8.** $5 \times 3 = 15$
9. $3 \times 3 = 9$

10. $6 \div 3 = 2$ **II.** $9 \div 3 = 3$ **12.** $30 \div 3 = 10$ **13.** $21 \div 3 = 7$
14. $27 \div 3 = 9$ **15.** $24 \div 3 = 8$ **16.** $3 \div 3 = 1$ **17.** $15 \div 3 = 5$
18. $12 \div 3 = 4$ **19.** $18 \div 3 = 6$

Explore
$1 \to 3$ $2 \to 6$ $3 \to 9$ $4 \to 12$
($5 \to 15$ – not possible because you would need two 5 cards)
$6 \to 18$ $7 \to 21$ $8 \to 24$ $9 \to 27$

Some of the possible answers beyond 3 x 10 are:
$12 \to 36$ $16 \to 48$ $18 \to 54$ $19 \to 57$ $21 \to 63$
$23 \to 69$ $26 \to 78$ $27 \to 81$ $29 \to 87$ and so on.
The highest possible answer using 2-digit numbers is $91 \to 273$

page 24
Fractions

I. $\frac{3}{4}$ **2.** $\frac{1}{4}$ **3.** $\frac{2}{3}$

4. $\frac{2}{4} = \frac{1}{2}$ **5.** $\frac{3}{6} = \frac{1}{2}$ **6.** $\frac{1}{3}$

7. $\frac{2}{5}$ **8.** $\frac{4}{6} = \frac{2}{3}$ **9.** $\frac{4}{6} = \frac{2}{3}$

10. $\frac{3}{8}$ **II.** red $\frac{3}{4}$ blue $\frac{1}{4}$ **12.** red $\frac{2}{5}$ blue $\frac{3}{5}$

13. red $\frac{1}{3}$ blue $\frac{2}{3}$ **14.** red $\frac{1}{6}$ blue $\frac{5}{6}$ **15.** red $\frac{3}{8}$ blue $\frac{5}{8}$

16. red $\frac{2}{3}$ blue $\frac{1}{3}$ **17.** red $\frac{3}{7}$ blue $\frac{4}{7}$

Number Textbook 2

page 25
Fractions

1. 3 coloured squares shown
2. 2 coloured squares shown
3. 3 coloured squares shown
4. 4 coloured squares shown
5. 3 coloured squares shown
6. 4 coloured squares shown
7. 6 coloured squares shown
8. 4 coloured squares shown
9. 5 coloured squares shown
10. 10 coloured squares shown

❷ 1. $\frac{5}{8}$ 2. $\frac{1}{3}$ 3. $\frac{1}{4}$ 4. $\frac{1}{5}$ 5. $\frac{3}{6}$ 6. $\frac{6}{10}$ 7. $\frac{3}{9}$ 8. $\frac{2}{6}$ 9. $\frac{3}{8}$ 10. $\frac{0}{10}$

Explore
Answers will vary.

---

page 26
Fractions

1. $\frac{1}{4}$ of 8 = 2
2. $\frac{1}{3}$ of 6 = 2
3. $\frac{3}{8}$ of 8 = 3
4. $\frac{2}{5}$ of 10 = 4
5. $\frac{5}{6}$ of 12 = 10
6. $\frac{3}{4}$ of 8 = 6
7. $\frac{2}{3}$ of 6 = 4
8. $\frac{4}{5}$ of 10 = 8
9. $\frac{1}{4}$ of 12 = 3
10. $\frac{2}{3}$ of 9 = 6
11. $\frac{1}{3}$ of 15 = 5 boys
12. $\frac{1}{4}$ of 20 = 5 raisins
13. $\frac{3}{5}$

---

page 27
Nearest ten

1. **a** 23   **b** 26   **c** 29
2. **d** 43   **e** 47   **f** 51   **g** 53   **h** 56
3. **i** 66   **j** 74   **k** 79   **l** 88   **m** 92

4. 16p → 20p
5. 39p → 40p
6. 44p → 40p
7. 47p → 50p
8. 34p → 30p
9. 27p → 30p

---

page 28
Nearest hundred

1. **a** 130   **b** 160   **c** 190
2. **d** 420   **e** 480   **f** 520   **g** 570
3. **h** 260   **i** 290   **j** 420   **k** 580

❷ **a** 100   **b** 200   **c** 200   **d** 400   **e** 500   **f** 500   **g** 600   **h** 300
i 300   **j** 800

Number Textbook 2

page 28 cont ...

4. 460g → 500g   5. 320g → 300g   6. 940g → 900g
7. 680g → 700g   8. 170g → 200g   9. 550g → 600g
10. 770g → 800g

---

page 29
Nearest hundred

1. 403 → 400 miles   2. 171 → 200 miles   3. 128 → 100 miles
4. 188 → 200 miles   5. 278 → 300 miles   6. 390 → 400 miles
7. 264 → 300 miles   8. 84 → 100 miles   9. 636 → 600 miles

Explore

347	349	374	379	394	397
437	439	473	479	493	497
734	739	743	749	793	794
934	937	943	947	973	974

---

page 30
Adding multiples of 5

$5 + 95 = 100$ $10 + 90 = 100$ $15 + 85 = 100$ $20 + 80 = 100$
$25 + 75 = 100$ $30 + 70 = 100$ $35 + 65 = 100$ $40 + 60 = 100$
$45 + 55 = 100$ $50 + 50 = 100$ $55 + 45 = 100$ $60 + 40 = 100$
$65 + 35 = 100$ $70 + 30 = 100$ $75 + 25 = 100$ $80 + 20 = 100$
$85 + 15 = 100$ $90 + 10 = 100$ $95 + 5 = 100$

1. 45p + 25p = 70p   2. 45p + 35p = 80p   3. 45p + 5p = 50p
4. 45p + 55p = 100p   5. 45p + 15p = 60p   6. 25p + 35p = 60p
7. 25p + 5p = 30p   8. 25p + 55p = 80p   9. 25p + 15p = 40p
10. 35p + 5p = 40p   11. 35p + 55p = 90p   12. 35p + 15p = 50p
13. 5p + 55p = 60p   14. 5p + 15p = 20p   15. 55p + 15p = 70p

---

page 31
Adding multiples of 5

1. 25 + 30 = 55   55 + 15 = 70cm   2. 35 + 50 = 85   85 + 15 = 100cm
3. 30 + 45 = 75   75 + 25 = 100cm   4. 25 + 25 = 50   50 + 15 = 65cm
5. 40 + 15 = 55   55 + 45 = 100cm   6. 15 + 15 = 30   30 + 15 = 45cm

Number Textbook 2
page 31 cont ...
Explore
It is possible to make all scores between 50 and 75 so there are 24 different scores.

---

page 32
Adding three 2-digit numbers

Jon 100	Ashley 45	Ben 115	Rosa 72	Ana 105
Sumi 120	Jenny 135	Tom 130	Pete 110	Aziz 60
Bec 105	Jodi 70	Nick 32	Jenny wins	

---

page 33
Taking away 10, 20, 30

1. 3:28 → 3:08
2. 6:46 → 6:26
3. 4:38 → 4:18
4. 8:57 → 8:37
5. 5:24 → 5:04
6. 10:31 → 10:11
7. 7:40 → 7:20
8. 3:29 → 3:09
9. 11:49 → 11:29
10. 2:48 → 2:28
11. 11:55 → 1:35
12. 3:46 → 3:26
13. 12:25 → 12:05

in	42	57	73	95	39	64	87	66
out	12	27	43	65	9	34	57	36

---

page 34
Taking away

1. 65p – 25p = 40p
2. 35p – 25p = 10p
3. 75p – 25p = 50p
4. 55p – 25p = 30p
5. 85p – 25p = 60p
6. 45p – 25p = 20p
7. 95p – 25p = 70p
8. 30p – 25p = 5p
9. 50p – 25p = 25p
10. 70p – 25p = 45p
11. 45p – 15p = 30p
12. 60p – 35p = 25p
13. 65p – 45p = 20p

Number Textbook 2

page 35
Taking away

I. $56p - 23p = 33p$ 2. $54p - 31p = 23p$ 3. $46p - 24p = 22p$
4. $28p - 15p = 13p$ 5. $42p - 41p = 1p$ 6. $96p - 65p = 31p$
7. $85p - 42p = 43p$ 8. $69p - 28p = 41p$ 9. $44p - 13p = 31p$

@ Answers will vary.

10. $49 - 32 = 17$ $49 - 23 = 26$ $49 - 14 = 35$ $68 - 32 = 36$ $68 - 23 = 45$
     $68 - 14 = 54$ $75 - 32 = 43$ $75 - 23 = 52$ $75 - 14 = 61$

page 36
Difference

I. $8 + 5 = 13$ 2. $19 + 3 = 22$ 3. $28 + 5 = 33$ 4. $39 + 5 = 44$
5. $48 + 5 = 53$ 6. $57 + 5 = 62$ 7. $68 + 5 = 73$ 8. $79 + 6 = 85$
9. $88 + 5 = 93$

10. $17 + 6 = 23$ $23 - 17 = 6$ II. $37 + 5 = 42$ $42 - 37 = 5$
12. $28 + 6 = 34$ $34 - 28 = 6$ 13. $57 + 4 = 61$ $61 - 57 = 4$

page 37
Difference

I. $18p + 5p = 23p$ $23p - 18p = 5p$ 2. $17p + 8p = 25p$ $25p - 17p = 8p$
3. $25p + 7p = 32p$ $32p - 25p = 7p$ 4. $37p + 7p = 44p$ $44p - 37p = 7p$
5. $28p + 6p = 34p$ $34p - 28p = 6p$ 6. $45p + 7p = 52p$ $52p - 45p = 7p$
7. $66p + 5p = 71p$ $71p - 66p = 5p$

8. $42 - 37 = 5$ 9. $34 - 29 = 5cm$ 10. $53 - 48 = 5$ miles
II. $65 - 57 = 8$ m

page 38
Difference

I. $53 - 48 = 5$ 2. $42 - 39 = 3$ 3. $33 - 28 = 5$ 4. $21 - 16 = 5$
5. $52 - 47 = 5$ 6. $22 - 18 = 4$ 7. $31 - 25 = 6$ 8. $54 - 45 = 9$
9. $61 - 58 = 3$ 10. $72 - 65 = 7$

Number Textbook 2
page 38 cont ...
Explore

$24 - 19 = 5$ $\quad$ $36 - 29 = 7$ $\quad$ $72 - 69 = 3$ $\quad$ $44 - 39 = 5$ $\quad$ $53 - 49 = 4$

$12 - 9 = 3$ $\quad$ $65 - 59 = 6$ $\quad$ $87 - 79 = 8$

The units digit increases by I.

@ Answers will vary.

---

page 39
Adding three numbers

I. $15 + 35 = 50$ $\quad$ $50 + 12 = 62$ $\qquad$ **2.** $25 + 25 = 50$ $\quad$ $50 + 13 = 63$

3. $15 + 15 = 30$ $\quad$ $30 + 14 = 44$ $\qquad$ **4.** $5 + 75 = 80$ $\quad$ $80 + 18 = 98$

5. $15 + 65 = 80$ $\quad$ $80 + 12 = 92$ $\qquad$ **6.** $35 + 35 = 70$ $\quad$ $70 + 21 = 91$

7. $45 + 15 = 60$ $\quad$ $60 + 16 = 76$ $\qquad$ **8.** $55 + 25 = 80$ $\quad$ $80 + 12 = 92$

9. $65 + 15 = 80$ $\quad$ $80 + 13 = 93$ $\qquad$ **10.** $75 + 14 = 89$ $\quad$ $89 + 25 = 114$

II. $28 + 5 = 33$ $\quad$ $33 + 65 = 98$ $\qquad$ **12.** $45 + 17 = 62$ $\quad$ $62 + 15 = 77$

$25 + 35 + 15 = 75$ $\qquad$ $25 + 35 + 45 = 105$ $\qquad$ $25 + 35 + 12 = 72$

$25 + 15 + 12 = 52$ $\qquad$ $25 + 15 + 45 = 85$ $\qquad$ $25 + 12 + 45 = 82$

$35 + 15 + 12 = 62$ $\qquad$ $35 + 15 + 45 = 95$ $\qquad$ $35 + 12 + 45 = 92$

$15 + 12 + 45 = 72$

@ 52 $\quad$ 62 $\quad$ 72 $\quad$ 72 $\quad$ 75 $\quad$ 82 $\quad$ 85 $\quad$ 92 $\quad$ 95 $\quad$ 105

---

page 40
Adding three numbers

I. $15 + 25 = 40$ $\quad$ $40 + 23 = 63$ $\qquad$ **2.** $5 + 24 = 29$ $\quad$ $29 + 55 = 84$

3. $35 + 33 = 68$ $\quad$ $68 + 25 = 93$ $\qquad$ **4.** $65 + 22 = 87$ $\quad$ $87 + 5 = 92$

5. $45 + 25 = 70$ $\quad$ $70 + 23 = 93$ $\qquad$ **6.** $15 + 35 = 50$ $\quad$ $50 + 42 = 92$

7. $5 + 61 = 66$ $\quad$ $66 + 25 = 91$ $\qquad$ **8.** $31 + 35 = 66$ $\quad$ $66 + 25 = 91$

9. $25 + 21 = 46$ $\quad$ $46 + 55 = 101$ $\qquad$ **10.** $32 + 25 = 57$ $\quad$ $57 + 45 = 102$

Explore
The star card can be any number between 10 and 54. The heart card can be any number between I and 45.
For example: i.e. heart I, star 54 <u>or</u> heart 45, star 10.

Number Textbook 2

page 41
Adding three numbers

1. $35 + 25 + 12 = 72$ **2.** $45 + 25 + 14 = 84$ **3.** $12 + 45 + 35 = 92$
4. $25 + 14 + 35 = 74$ **5.** $45 + 25 + 17 = 87$ **6.** $25 + 17 + 35 = 77$

ℯ All the following are possible.

$12 + 25 + 14 = 51$	$12 + 25 + 45 = 82$	$12 + 25 + 17 = 54$
$12 + 14 + 17 = 43$	$12 + 14 + 35 = 61$	$12 + 14 + 45 = 71$
$12 + 45 + 17 = 74$	$12 + 17 + 35 = 64$	$25 + 14 + 17 = 56$
$25 + 45 + 35 = 105$	$14 + 45 + 17 = 76$	$14 + 45 + 35 = 94$
$14 + 17 + 35 = 66$	$45 + 17 + 35 = 97$	

highest 105   lowest 43

7. $25 + 35 + 24 = 84$ **8.** $55 + 15 + 28 = 98$ **9.** $65 + 25 + 18 = 108$

---

page 42
Adding to 3-digit numbers

1. $232 + 8 = 240$ **2.** $143 + 7 = 150$ **3.** $154 + 6 = 160$
4. $378 + 2 = 380$ **5.** $186 + 4 = 190$ **6.** $275 + 5 = 280$
7. $131 + 9 = 140$ **8.** $252 + 8 = 260$ **9.** $196 + 4 = 200$
10. $333 + 7 = 340$ **11.** $146 + 20 = 166$ **12.** $255 + 30 = 285$
13. $374 + 20 = 394$ **14.** $417 + 40 = 457$ **15.** $525 + 50 = 575$

---

page 43
Adding to 3-digit numbers

1. $126p + 6p = 132p$ **2.** $164p + 6p = 170p$ **3.** $128p + 6p = 134p$
4. $137p + 6p = 143p$ **5.** $155p + 6p = 161p$ **6.** $228p + 6p = 234p$
7. $177p + 6p = 183p$ **8.** $293p + 6p = 299p$

ℯ **1.** 142p **2.** 180p **3.** 144p **4.** 153p **5.** 171p **6.** 244p **7.** 193p **8.** 309p

9. $155p + 20p = 175p$ **10.** $146 + 9 = 155$ stickers
11. $246 + 8 = 254$ stamps

---

page 44
Adding to 3-digit numbers

1. $124 + 22 = 146$ **2.** $136 + 23 = 159$ **3.** $245 + 25 = 270$
4. $150 + 36 = 186$ **5.** $346 + 31 = 377$ **6.** $444 + 44 = 488$
7. $285 + 15 = 300$ **8.** $625 + 25 = 650$ **9.** $348 + 51 = 399$
10. $255 + 31 = 286$ **11.** $152 + 34 = 186$ **12.** $176 + 13 = 189$
13. $224 + 24 = 248$

Number Textbook 2

page 44 cont ...

Explore

There are 24 possible additions:

123 + 4 = 127	124 + 3 = 127	134 + 2 = 136	234 + 1 = 235
132 + 4 = 136	142 + 3 = 145	143 + 2 = 145	243 + 1 = 244
213 + 4 = 217	214 + 3 = 217	314 + 2 = 316	324 + 1 = 325
231 + 4 = 235	241 + 3 = 244	341 + 2 = 343	342 + 1 = 343
312 + 4 = 316	412 + 3 = 415	413 + 2 = 415	423 + 1 = 424
321 + 4 = 325	421 + 3 = 424	431 + 2 = 433	432 + 1 = 433

@ smallest total 127
largest total 433

---

page 45

Adding multiples of one hundred

Addition/subtraction **N35**

1. 241 + 300 = 541	2. 366 + 400 = 766	3. 414 + 500 = 914
4. 179 + 500 = 679	5. 196 + 200 = 396	6. 344 + 600 = 944
7. 374 + 400 = 774	8. 299 + 300 = 599	9. 458 + 500 = 958
10. 213 + 200 = 413	11. 421 + 400 = 821	12. 444 + 500 = 944
13. 342 + 250 = 592	14. 431 + 250 = 681	15. 125 + 250 = 375
16. 616 + 250 = 866	17. 249 + 250 = 499	18. 750 + 250 = 1000

---

page 46

Adding two 3-digit numbers

Addition/subtraction **N35**

1. 423 + 200 + 6 = 629	2. 555 + 100 + 4 = 659
3. 324 + 200 + 2 = 526	4. 412 + 300 + 7 = 719
5. 821 + 100 + 8 = 929	6. 457 + 400 + 1 = 858
7. 221 + 500 + 8 = 729	8. 644 + 300 + 3 = 947

9. All the following are possible.

2.12 + 4.24 = £6.36

2.12 + 3.42 = £5.54	2.12 + 1.10 = £3.22	2.12 + 2.10 = £4.22	2.12 + 3.20 = £5.32
4.24 + 3.42 = £7.66	4.24 + 1.10 = £5.34	4.24 + 2.10 = £6.34	4.24 + 3.20 = £7.44
	3.42 + 1.10 = £4.52	3.42 + 2.10 = £5.52	3.42 + 3.20 = £6.62
		1.10 + 2.10 = £3.20	1.10 + 3.20 = £4.30
			2.10 + 3.20 = £5.30

Number Textbook 2

page 47
Subtracting 3-digit numbers

1. $417 - 110 = 307$	**2.** $667 - 220 = 447$	**3.** $426 - 210 = 216$
4. $576 - 240 = 336$	**5.** $744 - 310 = 434$	**6.** $352 - 120 = 232$
7. $474 - 150 = 324$	**8.** $691 - 230 = 461$	**9.** $535 - 320 = 215$
10. $642 - 210 = 432$ m	**11.** $731 - 320 = 411$ m	**12.** $543 - 130 = 413$ m
13. $422 - 210 = 212$ m	**14.** $311 - 110 = 201$ m	

page 48
Odd and even

1. 2	4	6	8	10	12						
2. 1	3	5	7	9	11	13	15	17	19	21	23
3. 34	36	38	40	42	44	46	48	50	52	54	56
4. 61	63	65	67	69	71	73	75	77	79	81	83
5. 70	72	74	76	78	80	82	84	86	88	90	92
6. 43	45	47	49	51	53	55	57	59	61	63	65

7. 43 odd	**8.** 24 even	**9.** 66 even	**10.** 35 odd
11. 71 odd	**12.** 82 even	**13.** 90 even	**14.** 55 odd
15. 64 even	**16.** 93 odd	**17.** 99 odd	

page 49
Odd and even

1–9. Answers will vary.

@ Answers will vary.

Explore

21	23	25	27	29
41	43	45	47	49
61	63	65	67	69
81	83	85	87	89 total 20

page 50
Odd and even

1. 74p even	**2.** 45p odd	**3.** 51p odd	**4.** 66p even
5. 92p even	**6.** 54p even	**7.** 63p odd	**8.** 59p odd
9. 78p even	**10.** 28p even	**11.** 33p odd	**12.** 97p odd
13. 81p odd			

Number Textbook 2
page 50 cont ...
Explore

20 & 1	18 & 3	16 & 5	14 & 7	12 & 9
10 & 11	8 & 13	6 & 15	4 & 17	2 & 19

---

page 51
Multiplication/division **N37**
Multiplying by 10 and 100

1. $4 \times 10 = 40$
2. $6 \times 10 = 60$
3. $3 \times 100 = 300$
4. $8 \times 10 = 80$
5. $4 \times 100 = 400$
6. $5 \times 10 = 50$
7. $5 \times 100 = 500$
8. $3 \times 100p = 300p$
9. $4 \times 100p = 400p$
10. $6 \times 100p = 600p$
11. $8 \times 100p = 800p$
12. $2 \times 100p = 200p$
13. $9 \times 100p = 900p$
14. $3 \times 100p = 300p$

---

page 52
Multiplication/division **N37**
Multiplying by 10 and 100

1. $6 \times 100 = 600cm$
2. $9 \times 100 = 900cm$
3. $2 \times 100 = 200cm$
4. $5 \times 100 = 500cm$
5. $3 \times 100 = 300cm$
6. $7 \times 100 = 700cm$
7. $8 \times 100 = 800cm$

8. $4 \times 10 = 40$
9. $2 \times 10 = 20$
10. $3 \times 100 = 300$
11. $7 \times 10 = 70$
12. $5 \times 100 = 500$
13. $9 \times 10 = 90$
14. $1 \times 100 = 100$
15. $6 \times 10 = 60$
16. $7 \times 100 = 700$
17. $1 \times 10 = 10$
18. $12 \times 10 = 120$
19. $4 \times 100 = 400$
20. $15 \times 10 = 150$

---

page 53
Multiplication/division **N37**
Multiplying by 10 and 100

1. $11 \times 10 = 110$
2. $19 \times 10 = 190$
3. $13 \times 10 = 130$
4. $10 \times 10 = 100$
5. $12 \times 10 = 120$
6. $14 \times 10 = 140$
7. $17 \times 10 = 170$
8. $20 \times 10 = 200$
9. $18 \times 10 = 180$

10. $5 \times 10 = 50$ questions
11. $12 \times 100 = 1200m$
12. $14 \times 10 = 140$ children

Number Textbook 2

page 54
Multiplying

Kim	2 x 30p = 60p	Toni	2 x 20p = 40p
	3 x 30p = 90p		3 x 20p = 60p
	4 x 30p = 120p = £1.20		4 x 20p = 80p
Sanjit	2 x 40p = 80p	Drew	2 x 10p = 20p
	3 x 40p = 120p = £1.20		3 x 10p = 30p
	4 x 40p = 160p = £1.60		4 x 10p = 40p
Tanya	2 x 50p = 100p = £1		
	3 x 50p = 150p = £1.50		
	4 x 50p = 200p = £2		

1. 4 x 20cm = 80cm
2. 3 x 30cm = 90cm
3. 4 x 40cm = 160cm
4. 5 x 30cm = 150cm
5. 3 x 20 cm = 60cm

---

page 55
Multiplying

1. 2 x 20p = 40p
2. 3 x 30p = 90p
3. 2 x 40p = 80p
4. 4 x 30p = 120p = £1.20
5. 3 x 50p = 150p = £1.50
6. 2 x 50p = 100p = £1
7. 3 x 40p = 120p = £1.20
8. 3 x 20p = 60p
9. 2 x 50p + 30p = 130p = £1.30
10. 2 x 20p + 2 x 40p = 120p = £1.20

Explore
There is only one of each number card, so the answers are as follows.

2 x 60 = 120	3 x 60 = 180	4 x 60 = 240	5 x 60 = 300
2 x 50 = 100	3 x 50 = 150	4 x 50 = 200	5 x 40 = 200
2 x 40 = 80	3 x 40 = 120	4 x 30 = 120	5 x 30 = 150
2 x 30 = 60	3 x 20 = 60	4 x 20 = 80	5 x 20 = 100

6 x 50 = 300
6 x 40 = 240
6 x 30 = 180
6 x 20 = 120

There are 9 different totals. There are 2 multiplications for each total (4 for the total 120).

Number Textbook 2

page 56
Multiplying

1. $2 \times 20 = 40$
2. $3 \times 50 = 150$
3. $4 \times 40 = 160$
4. $6 \times 20 = 120$
5. $5 \times 50 = 250$
6. $3 \times 50 + 3 \times 40 = 270$

7. $3 \times 30 = 90$
8. $4 \times 30 = 120$
9. $5 \times 20 = 100$
10. $2 \times 30 = 60$
11. $4 \times 50 = 200$
12. $4 \times 40 = 160$
13. $2 \times 50 = 100$
14. $3 \times 10 = 30$ buttons
15. $4 \times 30 = 120$ days
16. $2 \times 40 = 80$m

---

page 57
Fours

4   8   12   16   20   24   28   32   36   40

1. $2 \times 4 = 8$
2. $7 \times 4 = 28$
3. $5 \times 4 = 20$
4. $6 \times 4 = 24$
5. $3 \times 4 = 12$
6. $1 \times 4 = 4$
7. $4 \times 4 = 16$
8. $8 \times 4 = 32$
9. $9 \times 4 = 36$
10. $1 \times 4 = 4$
11. $6 \times 4 = 24$
12. $9 \times 4 = 36$
13. $2 \times 4 = 8$
14. $10 \times 4 = 40$
15. $7 \times 4 = 28$
16. $4 \times 4 = 16$
17. $3 \times 4 = 12$
18. $5 \times 4 = 20$
19. $8 \times 4 = 32$

@ $1 \times 4 = 4$ $2 \times 4 = 8$ $3 \times 4 = 12$ $4 \times 4 = 16$ $5 \times 4 = 20$
$6 \times 4 = 24$ $7 \times 4 = 28$ $8 \times 4 = 32$ $9 \times 4 = 36$ $10 \times 4 = 40$

---

page 58
Fours

1. $3 \times 4 = 12$ $12 \div 4 = 3$
2. $5 \times 4 = 20$ $20 \div 4 = 5$
3. $1 \times 4 = 4$ $4 \div 4 = 1$
4. $4 \times 4 = 16$ $16 \div 4 = 4$
5. $7 \times 4 = 28$ $28 \div 4 = 7$
6. $10 \times 4 = 40$ $40 \div 4 = 10$
7. $2 \times 4 = 8$ $8 \div 4 = 2$
8. $6 \times 4 = 24$ $24 \div 4 = 6$
9. $9 \times 4 = 36$ $36 \div 4 = 9$
10. $8 \times 4 = 32$ $32 \div 4 = 8$

11. $3 \times 4 = 12$
12. $2 \times 4 = 8$
13. $6 \times 4 = 24$
14. $20 \div 4 = 5$
15. $16 \div 4 = 4$
16. $8 \times 4 = 32$
17. $4 \div 4 = 1$
18. $7 \times 4 = 28$
19. $9 \times 4 = 36$
20. $40 \div 4 = 10$

Number Textbook 2

page 59
Fours

I. $28 \div 4 = 7m$ **2.** $16 \div 4 = 4m$ **3.** $12 \div 4 = 3m$ **4.** $32 \div 4 = 8m$
5. $4 \div 4 = 1m$ **6.** $20 \div 4 = 5m$ **7.** $24 \div 4 = 6m$

Explore
The numbers coloured should all be in the x4 table.
Patterns will vary.

---

page 60
Dividing

I. $12 \div 4 = 3$ **2.** $10 \div 5 = 2$ **3.** $12 \div 3 = 4$ **4.** $10 \div 2 = 5$
5. $14 \div 2 = 7$ **6.** $8 \div 4 = 2$ **7.** $15 \div 5 = 3$ **8.** $9 \div 3 = 3$
9. $21 \div 7 = 3$ **10.** $30 \div 6 = 5$

⊘ **I.** $3 \times 4 = 12$ **2.** $2 \times 5 = 10$ **3.** $4 \times 3 = 12$ **4.** $5 \times 2 = 10$ **5.** $7 \times 2 = 14$
   6. $2 \times 4 = 8$ **7.** $3 \times 5 = 15$ **8.** $3 \times 3 = 9$ **9.** $3 \times 7 = 21$ **10.** $5 \times 6 = 30$

II. $14 \div 2 = 7$ **12.** $8 \div 4 = 2$ **13.** $10 \div 5 = 2$ **14.** $16 \div 2 = 8$
15. $20 \div 10 = 2$ **16.** $15 \div 5 = 3$ **17.** $21 \div 3 = 7$ **18.** $16 \div 4 = 4$
19. $30 \div 10 = 3$

---

page 61
Remainders

I. $14 \div 3 = 4\ r2$ **2.** $14 \div 4 = 3\ r2$ **3.** $14 \div 2 = 7$ **4.** $14 \div 5 = 2\ r4$
5. $18 \div 5 = 3\ r3$ **6.** $18 \div 2 = 9$ **7.** $18 \div 4 = 4\ r2$ **8.** $18 \div 3 = 6$
9. $18 \div 6 = 3$ **10.** $18 \div 8 = 2\ r2$

Explore
$4 = 1 \times 3\ r1$ $8 = 2 \times 3\ r2$ $12 = 4 \times 3$ $16 = 5 \times 3\ r1$ $20 = 6 \times 3\ r2$
$24 = 8 \times 3$ $28 = 9 \times 3\ r1$ $32 = 10 \times 3\ r2$ $36 = 12 \times 3$ $40 = 13 \times 3\ r1$

$5 = 1 \times 3\ r2$ $0 = 3 \times 3\ r1$ $15 = 5 \times 3$ $20 = 6 \times 3\ r2$ $25 = 8 \times 3\ r1$ $30 = 10 \times 3$ etc
$6 = 2 \times 3$ $12 = 4 \times 3$ $18 = 6 \times 3$ $24 = 8 \times 3$ etc

Number Textbook 2

page 62
Remainders

1. $22 \div 4 = 5$ r2   6 cars
2. $24 \div 4 = 6$ 6 cars
3. $18 \div 4 = 4$ r2   5 cars
4. $20 \div 4 = 5$ 5 cars
5. $17 \div 4 = 4$ r1   5 cars
6. $13 \div 4 = 3$ r1   4 cars

7. $17 \div 3 = 5$ r2   5 tickets
8. $21 \div 3 = 7$ 7 tickets
9. $31 \div 3 = 10$ r1   10 tickets
10. $27 \div 5 = 5$ r2   5 teams
11. $27 \div 4 = 6$ r3   6 stickers $9 \times 4 = 36$p
12. $3 \times 5 = 15$ $17 - 15 = 2$ left

---

page 63
Matching fractions

1. $\frac{2}{6}$
2. $\frac{1}{2}$
3. $\frac{1}{3}$
4. $\frac{2}{4}$
5. $\frac{5}{6}$

6. $\frac{3}{4}$
7. $\frac{1}{4}$
8. $\frac{2}{3}$
9. $\frac{2}{8}$
10. $\frac{4}{6}$

🄮   1 and 3, 2 and 4, 8 and 10, 7 and 9

11. $\frac{2}{8} = \frac{1}{4}$
12. $\frac{4}{8} = \frac{1}{2}$
13. $\frac{4}{8} = \frac{1}{2}$
14. $1 = \frac{8}{8}$
15. $\frac{6}{8} = \frac{3}{4}$

16. $\frac{4}{8} = \frac{1}{2}$
17. $\frac{2}{8} = \frac{1}{4}$
18. $\frac{6}{8} = \frac{3}{4}$
19. $\frac{2}{8} = \frac{1}{4}$

---

page 64
Matching fractions

1. $\frac{1}{2} = \frac{2}{4}$
2. $\frac{4}{4} = 1$
3. $\frac{2}{8} = \frac{1}{4}$
4. $\frac{3}{4} = \frac{6}{8}$
5. $\frac{5}{10} = \frac{1}{2}$

6. $\frac{2}{3} = \frac{4}{6}$
7. $\frac{4}{8} = \frac{1}{2}$
8. $\frac{1}{3} = \frac{2}{6}$
9. $\frac{3}{5} = \frac{6}{10}$

10. $a = \frac{1}{4}$ $b = \frac{1}{2}$ $c = \frac{3}{4}$

11. $d = \frac{1}{8}$ $e = \frac{3}{8}$ $f = \frac{5}{8}$ $g = \frac{7}{8}$

12. $h = \frac{1}{6}$ $i = \frac{2}{6} = \frac{1}{3}$ $j = \frac{4}{6} = \frac{2}{3}$ $k = \frac{5}{6}$

---

page 65
Matching fractions

1. $\frac{1}{4} = \frac{2}{8}$
2. $\frac{1}{2} = \frac{2}{4} = \frac{4}{8}$
3. $\frac{3}{4} = \frac{6}{8}$
4. $\frac{1}{3} = \frac{2}{6}$
5. $\frac{1}{2} = \frac{3}{6}$

6. $\frac{2}{3} = \frac{4}{6}$
7. $\frac{1}{2} = \frac{4}{8}$ $\frac{2}{6} = \frac{1}{3}$ $\frac{2}{8} = \frac{1}{4}$ $\frac{4}{6} = \frac{2}{3}$ $\frac{3}{4} = \frac{6}{8}$

Number Textbook 2

page 66
Adding two 3-digit numbers

1. 243 + 100 = 343	343 + 9 = 352	**2.** 464 + 200 = 664	664 + 7 = 671
3. 555 + 300 = 855	855 + 6 = 861	**4.** 352 + 200 = 552	552 + 8 = 560
5. 277 + 100 = 377	377 + 4 = 381	**6.** 388 + 400 = 788	788 + 4 = 792
7. 626 + 200 = 826	826 + 5 = 831	**8.** 724 + 100 = 824	824 + 8 = 832
9. 357 + 300 = 657	657 + 3 = 660	**10.** 485 + 300 = 785	785 + 7 = 792
11. 538 + 200 = 738	738 + 3 = 741	**12.** 612 + 200 = 812	812 + 12 = 824

13. 436 + 160 = 596   **14.** 545 + 140 = 685   **15.** 438 + 240 = 678
16. 664 + 320 = 984   **17.** 527 + 250 = 777

---

page 67
Adding two 3-digit numbers

1. 427 + 231 = 658   **2.** 346 + 323 = 669   **3.** 251 + 317 = 568
4. 511 + 488 = 999   **5.** 104 + 125 = 229   **6.** 416 + 382 = 798
7. 383 + 415 = 798   **8.** 220 + 219 = 439   **9.** 191 + 308 = 499
10. 281 + 311 = 592

Explore
There are many possible answers, for example:
401 + 104 = 505   402 + 204 = 606   403 + 304 = 707
412 + 214 = 626   421 + 124 = 545   423 + 324 = 747
432 + 234 = 666   413 + 314 = 727   431 + 134 = 565 and so on.
The first and last digit in the total is always the same. The middle digit is always even.

---

page 68
Adding two 3-digit numbers

1–10. All the following are possible.
£241 + £353 = £594   £241 + £342 = £583   £241 + £512 = £753
£241 + £425 = £666   £241 + £134 = £375   £241 + £430 = £671
£353 + £342 = £695   £353 + £512 = £865   £353 + £425 = £778
£353 + £134 = £487   £353 + £430 = £783
£342 + £512 = £854   £342 + £425 = £767   £342 + £134 = £476
£342 + £430 = £772
£512 + £425 = £937   £512 + £134 = £646   £512 + £430 = £942
£425 + £134 = £559   £425 + £430 = £855
£134 + £430 = £564

Number Textbook 2
page 68 cont ...

@   £241 + £255 = £496   £353 + £255 = £608   £342 + £255 = £597
    £512 + £255 = £767   £425 + £255 = £680   £134 + £255 = £389
    £430 + £255 = £685

11. 428 + 216 = 644 pieces
12. 134 + 219 = 353m
13. 532 + 144 = 676 rows

---

page 69
Adding 3-digit numbers

   1. 175 + 237 = 412   2. 325 + 486 = 811   3. 364 + 168 = 532
   4. 291 + 619 = 910   5. 724 + 156 = 880   6. 444 + 459 = 903
   7. 247 + 647 = 894   8. 195 + 545 = 740   9. 459 + 158 = 617
  10. 267 + 544 = 811   11. 316 + 499 = 815   12. 239 + 566 = 805

---

page 70
Adding 2-digit and 3-digit numbers

   1. 127 + 158 = 285g   2. 248 + 158 = 406g   3. 347 + 158 = 505g
   4. 384 + 158 = 542g   5. 492 + 158 = 650g   6. 559 + 158 = 717g
   7. 649 + 158 = 807g   8. 199 + 158 = 357g   9. 632 + 158 = 790g
  10. 408 + 158 = 566g

Any of the following answers.
£1.29 + 82p = £2.11   £1.29 + 77p = £2.06   £1.29 + 69p = £1.98
£1.46 + 82p = £2.28   £1.46 + 77p = £2.23   £1.46 + 69p = £2.15
£1.55 + 82p = £2.37   £1.55 + 77p = £2.32   £1.55 + 69p = £2.24
@ £2.37

---

page 71
Adding 3-digit numbers

   1. 468 + 273 = 741g   2. 137 + 247 = 384g   3. 683 + 158 = 841g
   4. 195 + 627 = 822g   5. 338 + 246 = 584g   6. 399 + 515 = 914g
   7. 434 + 186 = 620g   8. 738 + 193 = 931g   9. 117 + 714 = 831g
  10. 326 + 495 = 821g

Explore
123 + 45 = 168   432 + 51 = 483   324 + 15 = 339

Number Textbook 2

page 72
Mixed problems

1. $2 \times 8 = 16$ $2 + 8 = 10$ cards 2,8
2. $7 + 8 = 15$ $8 - 7 = 1$ cards 7,8
3. $16 \div 2 = 8$ $16 + 8 = 24$ cards 8,16
4. $22 + 11 = 33$ cards 11,22
5. $5p + 10p + 20p = 35p$
6. $£1.70 + 50p + 25p + 55p = £3$
7. $2 \times 20p = 40p$, $3 \times 30p = 90p$, $40p + 90p + 5p = 135p = £1.35$
8. $5p + 10p + 20p + 50p = 85p$, $2 \times 85p = 170p = £1.70$

Shape, Data and Measures

Length **M1**

page 3
Centimetres (cm)

I–7. Estimates and lengths will vary.

8. 4cm	**9.** 7cm	**10.** 6cm	**II.** Icm

---

page 4
Metres (m)

Length **M1**

I. 2m   **2.** 2m   **3.** 3m   **4.** $1\frac{1}{2}$ m   **5.** Im
6. 5m   **7.** 2m

8–13. Answers will vary.

---

page 5
Metres (m)

Length **M1**

I. Ikm = 1000m   **2.** Ikm 100m = 1100m   **3.** Ikm 500m = 1500m
4. 2km = 2000m   **5.** 2km 500m = 2500m   **6.** Ikm 700m = 1700m
7. Ikm 400m = 1400m   **8.** 3km = 3000m   **9.** 2km 100m = 2100m
10. 10km = 10 000m

Explore
beach and back 2 x 750m = 1500m = 1.5km
video shop and back 2 x 500m = 1000m = 1 km

---

page 6
Centimetres (cm) and metres (m)

Length **M2**

I. 3m = 300cm   **2.** Im = 100cm   **3.** 2m 30cm = 230cm
4. Im 40cm = 140cm   **5.** 3m 25cm = 325cm   **6.** Im 50cm = 150cm
7. 2m 15cm = 215cm   **8.** 2m 33cm = 233cm   **9.** Im 5cm = 105cm

🍂 Answers will vary.

10. 248cm = 2m 48cm   **II.** 292cm = 2m 92cm   **12.** 475cm = 4m 75cm
13. 332cm = 3m 32cm   **14.** 127cm = Im 27cm   **15.** 255cm = 2m 55cm
16. 304cm = 3m 4cm   **17.** 150cm = Im 50cm   **18.** IIIcm = Im IIcm

🍂 **10.** 2.48m **II.** 2.92m **12.** 4.75m **13.** 3.32m **14.** 1.27m **15.** 2.55m
   16. 3.04m **17.** 1.50m **18.** 1.11m

Shape, Data and Measures

page 7
Centimetres (cm) and metres (m)

1. 3m 15cm + 25cm = 3m 40cm
2. 4m 25cm + 25cm = 4m 50cm
3. 3m 70cm + 25cm = 3m 95cm
4. 3m 25cm + 25cm = 3m 50cm
5. 4m 75cm + 25cm = 5m
6. 3m 80cm + 25cm = 4m 5cm
7. 4m 35cm + 25cm = 4m 60cm
8. 4m 5cm + 25cm = 4m 30cm

Explore
The puppy has grown 40 cm (55 cm – 15 cm).
He could be any of the following ages:
4 weeks (40 cm ÷ 4 = 10 cm per week)
5 weeks (40 cm ÷ 5 = 8 cm per week)
8 weeks (40 cm ÷ 8 = 5 cm per week).

page 8
Centimetres (cm) and metres (m)

1. 1m 22cm – 1m 2cm = 20cm
2. 1m 30cm – 1m 10cm = 20cm
3. 1m 10cm – 1m = 10cm
4. 1m 15cm – 95cm = 20cm
5. 1m 20cm – 90cm = 30cm
6. 1m 30cm – 95cm = 35cm
7. 1m 25cm – 85cm = 40cm
8. 5 x 1km 10m = 5km 50m
9. 4 x 1m 20cm = 4m 80cm   10m – 4m 80cm = 5m 20cm left
10. 1m 30cm + 1m 15cm = 2m 45cm

page 9
Quarter past, half past, quarter to

1. a and j
2. b and i
3. c and n
4. d and k
5. e and p
6. f and m
7. g and l
8. h and o

page 10
5 minutes

1. 30 mins
2. 5 mins
3. 15 mins
4. 10 mins
5. 20 mins
6. 20 mins
7. 35 mins
8. 40 mins
9. 10 mins
10. 30 mins
11. 10 mins
12. 15 mins
13. 30 mins
14. 40 mins

Shape, Data and Measures

page 11
5 minutes

1. 8:45	8:50	8:55	9:00			
2. 9:35	9:40	9:45	9:50	9:55	10:00	
3. 6:05	6:10	6:15	6:20	6:25	6:30	6:35
6:40	6:45	6:50	6:55	7:00		

4. 10:45 or quarter to 11   **5.** 30 mins or $\frac{1}{2}$ hr   **6.** 5:45 or quarter to 6

---

page 12
5 minutes

1. ten past 6	**2.** quarter past 10	**3.** ten past 8
4. twenty past 11	**5.** half past 5	**6.** twenty-five past 7
7. twenty-five to 4	**8.** twenty past 1	**9.** quarter to 3
10. five to 5	**11.** twenty to 10	**12.** ten to 11
13. quarter past 8	**14.** 5 mins	**15.** 30 mins
16. 90 mins	**17.** 155 mins	

---

page 13
5 minutes

1. five past 5	**2.** twenty five to 3	**3.** quarter past 8
4. twenty-five past 3	**5.** ten to 7	**6.** twenty past 5
7. ten past 10	**8.** quarter to 8	**9.** twenty to 2
10. five to 5	**11.** half past 9	**12.** ten past 11
13. five past 12		

14. 5:15	**15.** 8:20	**16.** 10:30	**17.** 7:05	**18.** 11:40
19. 6:10	**20.** 8:50	**21.** 2:55	**22.** 1:40	**23.** 3:35

---

page 14
5 minutes

1. ten to 3	**2.** twenty-five past 6	**3.** twenty-five to 11
4. five to 5	**5.** quarter past 8	**6.** twenty past 9
7. five to 6	**8.** 12 o'clock	**9.** five past 1

1. twenty past 2	**2.** five to 6	**3.** five past 10
4. twenty-five past 4	**5.** quarter to 8	**6.** ten to 9
7. twenty-five past 5	**8.** half past 11	**9.** twenty-five to 1

Shape, Data and Measures

page 14 cont ...

10. ends 6:40	**11.** ends 6:05	**12.** ends 8:20
13. ends 4:15	**14.** ends 9:50	**15.** ends 1:10
16. ends 3:00		

---

page 15   Capacity **M5**
Millilitres (ml)

1. less than 1 litre	**2.** less than 1 litre	**3.** more than 1 litre
4. less than 1 litre	**5.** less than 1 litre	**6.** equal to 1 litre
7. more than 1 litre	**8.** more than 1 litre	**9.** less than 1 litre

10. 15 ml   **11.** 250 ml   **12.** 1000 ml   **13.** 300 ml   **14.** 300 ml

---

page 16   Capacity **M5**
Millilitres (ml)

1. 300 ml   **2.** 500 ml   **3.** 800 ml   **4.** 100 ml   **5.** 600 ml
6. 200 ml   **7.** 700 ml   **8.** 400 ml   **9.** 1000 ml

10. 1 l = 1000 ml   **11.** $3 l + \frac{1}{2} l = 3500$ ml   **12.** 2 l = 2000 ml

13. $2 l + \frac{1}{2} l = 2500$ ml   **14.** $\frac{1}{2} l = 500$ ml   **15.** 3 l = 3000 ml

16. 6 l = 6000 ml   **17.** 4 l = 4000 ml

18. $2 l + \frac{1}{2} l + \frac{1}{2} l = 3000$ ml

---

page 17   Capacity **M5**
Millilitres (ml)

50 ml	200 ml	$\frac{1}{2}$ l	650 ml	900 ml	1 l
1500 ml	2 l	2500 ml	3000 ml	4 l	

1. 1 l − 350 ml = 650 ml   **2.** 500 ml ÷ 5 ml = 100 spoonfuls
3. 1000 ml ÷ 150 ml = 6 r 100 ml   6 full cupfuls

Explore   Larger volume when longer side forms circumference of cylinder.

Shape, Data and Measures

page 18
Hours and minutes

1. I hour = 60 minutes
2. I hour 30 minutes = 90 minutes
3. I hour 20 minutes = 80 minutes
4. I hour 50 minutes = 110 minutes
5. 2 hours = 120 minutes
6. I hour 40 minutes = 100 minutes
7. I hour 45 minutes = 105 minutes
8. 2 hours 10 minutes = 130 minutes
9. I hour 55 minutes = 115 minutes
10. I hour 10 minutes = 70 minutes

11. 75 minutes = I hour 15 minutes
12. 85 minutes = I hour 25 minutes
13. 90 minutes = I hour 30 minutes
14. 100 minutes = I hour 40 minutes
15. 80 minutes = I hour 20 minutes
16. 120 minutes = 2 hours
17. 130 minutes = 2 hours 10 minutes
18. 95 minutes = I hour 35 minutes
19. 180 minutes = 3 hours

page 19
Days and hours

1. Monday
2. Tuesday
3. Wednesday
4. Thursday
5. Friday
6. Saturday
7. Sunday

8. more than I day
9. less than 2 days
10. less than 3 days
11. more than 2 days
12. less than I day
13. more than 3 days
14. less than 4 days

page 20
Seconds and minutes

1. I minute 5 seconds
2. I minute 15 seconds
3. I minute 25 seconds
4. I minute 50 seconds
5. 2 minutes
6. I minute 30 seconds
7. 2 minutes 10 seconds
8. 60 x 60 x 24 = 86400 seconds
9. 24 x 7 = 168 hours
10. 60 x 24 x 2 = 2880 minutes

@ Answers will vary.

page 21
Grams (g) and kilograms (kg)

1–7. Answers will vary.

8. 300g
9. 200g
10. 50g
11. 250g
12. 100g
13. 100g
14. 200g

Shape, Data and Measures

page 22
Grams (g) and kilograms (kg)

1. 4 kg 500g
2. 1 kg 700g
3. 7 kg 500g
4. 300g
5. 2 kg 500g
6. 3 kg 400g
7. 2 kg 400g
8. 2
9. 5
10. 10
11. 4
12. 20

Explore There are many possible combinations.

page 23
Grams (g) and kilograms (kg)

1. 1 kg = 1000g
2. $\frac{1}{2}$ kg = 500g
3. 2 kg = 2000g
4. $1\frac{1}{2}$ kg = 1500g
5. 1 kg 300g = 1300g
6. $\frac{3}{4}$ kg = 750g
7. 5 kg = 5000g

Problems

6 x 40g = 240g   1000g ÷ 50g = 20 tomatoes   10 x $\frac{1}{2}$ kg = 5 kg

page 24
Months

January	February	March	April
May	June	July	August
September	October	November	December

1. 16 months = 1 year 4 months
2. 26 months = 2 years 2 months
3. 30 months = 2 years 6 months
4. 20 months = 1 year 8 months
5. 23 months = 1 year 11 months

page 25
Days

1. 2 weeks = 14 days
2. 6 weeks = 42 days
3. 2 weeks 2 days = 16 days
4. 2 weeks 5 days = 19 days
5. 2 weeks 1 day = 15 days
6. 2 weeks 6 days = 20 days
7. 3 weeks = 21 days
8. 1 week 5 days = 12 days

9. 7 days
10. 10 days
11. 11 days
12. 15 days
13. 14 days
14. 13 days
15. 12 days

Shape, Data and Measures

page 26
Days, months and years

1. 7 days in 1 week
2. 28 days in 4 weeks
3. 10 years in a decade
4. 24 months in 2 years
5. 52 weeks in a year
6. 365 days in a year (not leap year)
7. 26 weeks in half a year
8. 104 weeks in 2 years
9. 120 months in a decade
10. 100 years in a century

Explore   Answers will vary.

---

page 27
2-d shape   S1

Shape names

1. 5 sides pentagon
2. 4 sides rectangle
3. 4 sides square
4. 3 sides triangle
5. 6 sides hexagon
6. 5 sides pentagon
7. 8 sides octagon
8. 3 sides triangle
9. 4 sides square
10. 8 sides octagon

Explore   Answers will vary.

---

page 28
2-d shape   **S1**

Quadrilaterals

1. yes
2. yes
3. hexagon
4. triangle
5. yes
6. yes
7. octagon
8. yes
9. pentagon
10. yes
11. octagon
12. octagon
13. pentagon
14. yes

---

page 29
2-d shape   S1

Shape names

1. hexagon
2. pentagon
3. rectangle
4. octagon
5. quadrilateral
6. hexagon
7. pentagon
8. triangle
9. pentagon

@ Answers will vary.

Explore   Answers will vary.

Shape, Data and Measures

page 30
Shape names
1. quadrilateral
2. pentagon
3. hexagon
4. pentagon
5. pentagon
6. hexagon

Explore   Answers will vary.

---

page 31
Lines of symmetry
1. yes 2. yes 3. no 4. yes 5. yes
6. no 7. yes 8. no 9. yes 10. no

❷ Answers will vary.

Bottom of page 31

 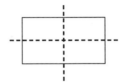

4 lines of symmetry   2 lines of symmetry

Shape, Data and Measures

page 32

Lines of symmetry

◉ Answers will vary.

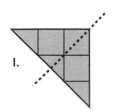

I.

2.
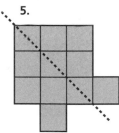

3.
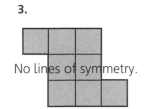
No lines of symmetry.

4.
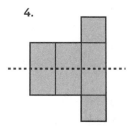

5.
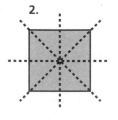

6.
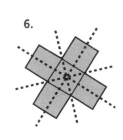

7.
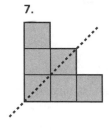

8.
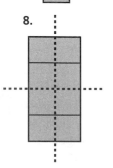

q.
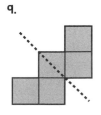

10.
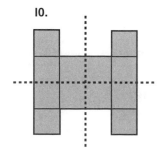

Shape, Data and Measures

page 33
Lines of symmetry

1. 1	**2.** 4	**3.** 3	**4.** 1
5. 4	**6.** 2	**7.** 4	**8.** 8
9. 4	**10.** 1		

Explore Answers will vary.

---

page 34
North, south, east, west

1. west	**2.** south	**3.** north	**4.** east	**5.** south
6. north	**7.** west	**8.** east	**9.** south	**10.** north

---

page 35
North, south, east, west

1. north	**2.** west	**3.** north	**4.** west	**5.** south
6. north	**7.** east	**8.** south	**9.** east	

---

page 36
North, south, east, west

1. west	**2.** north	**3.** east	**4.** north	**5.** west
6. south	**7.** west	**8.** south	**9.** west	**10.** south
11. west	**12.** north	**13.** east	**14.** north	**15.** east
16. north				

Explore Answers will vary.

---

page 37
Prisms

1. yes – square	**2.** yes – triangle	**3.** no	**4.** yes – hexagon
5. no	**6.** yes – square	**7.** no	**8.** yes – pentagon
9. no			

Shape, Data and Measures

page 38
Prisms

1. 2 triangles, 3 rectangles
2. 2 squares, 4 rectangles
3. 2 triangles, 3 rectangles
4. 2 hexagons, 6 rectangles
5. 6 rectangles
6. 6 squares
7. 2 octagons, 8 rectangles
8. 2 hexagons, 6 rectangles
9. 2 pentagons, 5 rectangles

---

page 39
Names of shapes

1. cube
2. cuboid
3. pyramid (either, triangular- or square-based)
4. cylinder
5. cone
6. triangular prism
7. square-based pyramid
8. cylinder (or prism)
9. sphere
10. cuboid
11. 4, 5, 8, 9
12. 1, 2, 10
13. 4, 8
14. 6
15. 1
16. 7 (some may include 3)
17. 5

---

page 40
Sorting shapes

Question	1	2	3	4	5	6	7	8	9	10
faces	6	6	3	5	3	3	6	8	3	6
vertices	8	8	0	5	0	0	8	12	0	8
edges	12	12	2	8	2	2	12	18	2	12

Explore Answers will vary.

---

page 41
Sorting shapes

shape	b	d	f	a	c	e
faces	4	5	5	6	7	8

Shape, Data and Measures
page 41 cont ...

shape	b	f	d	a	c	e
edges	6	8	9	12	15	18

shape	b	f	d	a	c	e
vertices	4	5	6	8	10	12

page 42
Sorting shapes

There are 29 prisms comprising:

6 cylinders   6 cubes   14 cuboids

Explore Answers will vary.

page 43
Right angles

I. no	**2.** yes	**3.** no	**4.** no	**5.** no	**6.** no
7. yes	**8.** no	**9.** yes	**10.** yes	**II.** no	**I2.** yes

I3. I right angle   **I4.** 2 right angles   **I5.** 4 right angles
I6. 2 right angles   **I7.** I right angle
I8. 2 right angles (or 3 if pupils count external one)
I9. 3 right angles

page 44
Right angle turns

I. yes	**2.** yes	**3.** no	**4.** no	**5.** no
6. yes	**7.** yes	**8.** no	**9.** yes	

I0. 3 right angles   @ anticlockwise   **II.** I right angle   @ anticlockwise
I2. I right angle   @ anticlockwise   **I3.** 2 right angles   @ clockwise
I4. 3 right angles   @ anticlockwise   **I5.** 2 right angles   @ clockwise
I6. 3 right angles   @ anticlockwise   **I7.** I right angle   @ clockwise
I8. 4 right angles   @ anticlockwise

Shape, Data and Measures

page 45
Right angle turns

I. 2 right angles		**2.** I right angle	
3. 3 right angles		**4.** 0 or 4 right angles	
5. 3 right angles		**6.** I right angle	
7. 2 right angles		**8.** 0 or 4 right angles	

page 46
Right angle turns

I. slide	**2.** climbing frame	**3.** slide	**4.** slide
5. roundabout	**6.** see-saw	**7.** slide	**8.** slide

page 47
Turning north, south, east or west

I. anticlockwise	west	**2.** clockwise	north
3. anticlockwise	north	**4.** clockwise	east
5. clockwise	south	**6.** anticlockwise	east
7. clockwise	west	**8.** anticlockwise	south

9. east	anticlockwise	north	anticlockwise	west		
10. north	clockwise	east	anticlockwise	north		
II. south	anticlockwise	east	clockwise	south		
12. north	anticlockwise	west	anticlockwise	south		
13. west	anticlockwise	south	clockwise	west		
14. south	anticlockwise	east	anticlockwise	north		
15. south	clockwise	west	anticlockwise	south		
16. west	anticlockwise	south	anticlockwise	east		
17. west	anticlockwise	south	anticlockwise	east	clockwise	south

page 48
Turning north, south, east or west

I. east to south		**2.** east to west	
3. north to east		**4.** west to south	
5. north to south		**6.** south to west	
7. north to south		**8.** south to east	
9. east to north			

Explore   Answers will vary.

Shape, Data and Measures

page 49
Turning north, south, east or west

1. south to east   anticlockwise
2. east to south   anticlockwise
3. north to south   anticlockwise
4. west to north   clockwise
5. south to south   anticlockwise
6. east to west   clockwise
7. north to west   clockwise
8. north to west   anticlockwise
9. west to east   clockwise

Explore   Answers will vary.

---

page 50
Position

1. column 3, row 3
2. column 2, row 1
3. column 4, row 2
4. column 4, row 4
5. column 2, row 4
6. column 2, row 2
7. column 4, row 1
8. column 1, row 3
9. column 1, row 2

10. Matt   11. Kirsty   12. Melita   13. Rory   14. Josie   15. Dan   16. Tom

---

page 51
Position

1. blue planet
2. yellow monster
3. blue spaceship
4. orange monster
5. red planet
6. orange spaceship
7. A4
8. A1
9. D5
10. E1
11. B4
12. B6

Answers are roughly:
- North → red monster
  yellow monster
  West → blue monster
  red spaceship
  East → red planet, blue planet
  South → blue spaceship, yellow spaceship

---

page 52
Position

1. 2A   2. 4B   3. 2D   4. 5B   5. 5A   6. 4D

7. chocolate   8. fudge   9. raisins   10. crisps   11. mints   12. lolly

Explore   Answers will vary.

Shape, Data and Measures

page 53
Tally charts

1. 5   **2.** 17   **3.** 9   **4.** 11   **5.** 3   **6.** ball
7. yoyo   **8.** transformer and ball   **9.** doll and yoyo

yoyo	6
dinosaur	10
transformer	8
ball	9
doll	7

page 54
Tally charts

Fruits in the picture

Fruits	Tallies
apples	JHT JHT
oranges	JHT JHT IIII
bananas	JHT JHT JHT JHT
mangoes	III
pineapples	JHT I
kiwis	JHT JHT IIII

Our favourite colours

Colour	Tallies	Total
Red	JHT JHT I	11
Blue	JHT JHT JHT JHT I	21
Yellow	JHT II	7
Purple	JHT JHT	10
Green	JHT IIII	9
Pink	IIII	4

page 55
Tally charts

Colour of caps

Colour	Tallies	Total
Orange	JHT JHT I	11
Red	JHT IIII	9
Yellow	JHT JHT JHT	15
Purple	JHT	5
Blue	JHT JHT JHT III	18
Pink	II	2

Explore – Answers will vary.

Shape, Data and Measures

page 56
Frequency tables

dice number	frequency
1	3
2	4
3	5
4	4
5	6
6	2

1. dice number 5　**2.** dice number 6　**3.** dice number 3　**4.** dice number 1
5. 7　　　　　　　**6.** 9　　　　　　**7.** 14　　　　　　**8.** 10

---

page 57
Frequency tables

card number	frequency
2	4
3	3
4	4
5	1
6	3
7	4
8	3

suit	frequency
hearts	7
diamonds	6
clubs	4
spades	5

1. 4s　　4　　　**2.** 6s　　3　　　**3.** 3s　　3　　　**4.** 5s　　1

5. 3 or more　18　　**6.** less than 5　11　　**7.** odd　　8
8. even　　14　　**9.** red　　13　　**10.** black　　9

Shape, Data and Measures

page 58
Frequency tables

1. football
2. swimming
3. cricket
4. roller blading
5. swimming
6. netball
7. football
8. roller blading

32 boys voted   34 girls voted

e

Sport	Frequency
Football	16
Running	4
Swimming	18
Roller blading	7
Cycling	3
Cricket	1
Rounders	6
Netball	4
Skating	7

Explore Answers will vary.

---

page 59
Bar graphs

1. football   9
2. cycling   6
3. swimming   10
4. rounders   5
5. tennis   9
6. running   3
7. swimming
8. running
9. football, rounders, swimming, tennis, cycling
10. rounders, running, cycling
11. 42 children voted

Explore Answers will vary.

Shape, Data and Measures

page 60
Bar graphs

superhero	frequency
computer kid	6
robo dog	9
brainy baby	5
galaxy girl	5

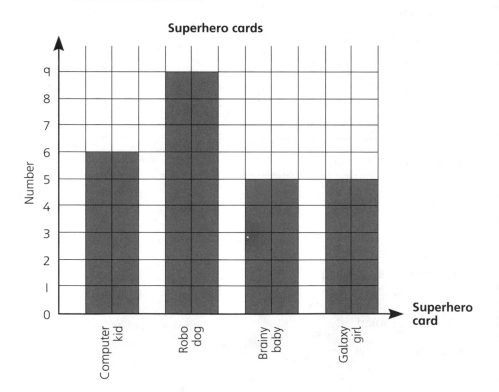

Superhero cards

Shape, Data and Measures

page 61
Bar graphs

coin	frequency
1p	7
2p	3
5p	1
20p	8
£1	5

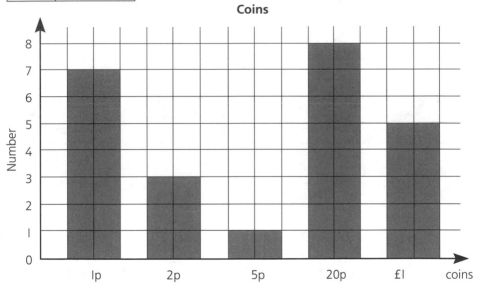

Coins

Explore Answers will vary.

page 62
Pictographs

1. strawberry
2. toffee
3. vanilla   5
4. toffee   3
5. strawberry   12
6. choc chip   7
7. minty   8

8. choc chip or toffee   10
9. minty or strawberry   20
10. 4 more
11. 2 more
12. 5 more
13. 9 more

35 children voted in total

Shape, Data and Measures

page 63
Pictographs

Type of ball	Balls			
tennis ball	◉	◉	◉	◉
football	◉	◉	(	
beach ball	◉	◉		
basketball	◉	◉	◉	(

Number of balls

Key
◉ = 2 balls
(= 1 ball

I. beach balls	4		**2.** tennis balls	8
3. basketballs	7		**4.** footballs	5
5. 16		**6.** 19	**7.** 2 more	**8.** 4 more

---

page 64
Pictographs

Type of cake	Cakes						
chocolate cake	🧁	🧁	🧁	◁			
jalebi	🧁	🧁	🧁	🧁	🧁	🧁	🧁
baklava	🧁	🧁	🧁	🧁	◁		
sponge cake	🧁	◁					
fruit cake	🧁	🧁	🧁	🧁			

Number of cakes

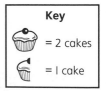

Key
🧁 = 2 cakes
◁ = 1 cake

I. jalebi		**2.** sponge cake	
3. fruit cakes	8	**4.** sponge cakes	3
5. jalebi	14	**6.** baklava	9
7. chocolate cakes	7		

Explore Answers will vary.

Photocopy Masters

page 1
Numbers to 1000

1. 3 x £1, 4 x 10p, 2 x 1p
2. 4 x £1, 6 x 10p, 9 x 1p
3. 2 x £1, 6 x 10p, 1 x 1p
4. 3 x £1, 4 x 10p, 7 x 1p
5. 1 x £1, 4 x 10p, 6 x 1p
6. 4 x £1, 3 x 10p, 1 x 1p
7. 1 x £1, 0 x 10p, 2 x 1p
8. 5 x £1, 5 x 10p, 5 x 1p

page 2
Hundreds, tens and units

1. 243
2. 256
3. 133
4. 222
5. 183
6. 360
7. 109
8. 404
9. 160
10. 111
11. 346

page 3
Hundreds, tens and units

1.	245	246	257	2.	308	309	310
3.	289	290	291	4.	970	971	972
5.	997	998	999	6.	110	111	112
7.	443	444	445	8.	239	240	241
9.	330	331	332	10.	888	889	890
11.	900	901	902	12.	666	667	668
13.	476	477	478	14.	561	562	563
15.	400	401	402				

page 4
3-digit numbers

Answers will vary.

page 5
Adding to 20

17 + 3   5 + 15   12 + 8   18 + 2   13 + 7   14 + 6   16 + 4   1 + 19   9 + 11

page 6
Adding to 20

1. 17 + 3 = 20
2. 13 + 7 = 20
3. 10 + 10 = 20
4. 12 + 8 = 20
5. 15 + 5 = 20
6. 16 + 4 = 20
7. 11 + 9 = 20
8. 19 + 1 = 20
9. 18 + 2 = 20
10. 14 + 6 = 20
11. 20 + 0 = 20

Photocopy Masters

page 7
Adding several numbers

1. $5 + 5 + 3 = 13$	**2.** $4 + 6 + 7 = 17$	**3.** $4 + 8 + 6 = 18$
4. $7 + 3 + 3 = 13$	**5.** $2 + 8 + 9 = 19$	**6.** $8 + 6 + 2 = 16$
7. $7 + 8 + 3 = 18$	**8.** $10 + 7 + 0 = 17$	**9.** $1 + 7 + 9 = 17$
10. $9 + 3 + 1 = 13$	**11.** $1 + 1 + 9 = 11$	**12.** $7 + 5 + 3 = 15$
13. $4 + 4 + 6 = 14$	**14.** $8 + 2 + 2 = 12$	**15.** $12 + 8 + 6 = 26$
16. $5 + 4 + 5 = 14$	**17.** $6 + 6 + 4 = 16$	**18.** $2 + 8 + 3 = 13$
19. $3 + 7 + 6 = 16$	**20.** $5 + 6 + 5 = 16$	

page 8
Adding

	top row	second row	
1.	34		
2.	42	20	22
3.	26	18	8
4.	35	22	13
5.	41	20	21
6.	41	21	20
7.	32	13	19
8.	37	20	17

9–14 Answers will vary.

page 9
Fewest coins

13p = 10p + 2p + 1p
16p = 10p + 5p + 1p
35p = 20p + 10p + 5p
42p = 20p + 20p + 2p
19p = 10p + 5p + 2p + 2p
27p = 20p + 5p + 2p
53p = 50p + 2p + 1p
67p = 50p + 10p + 5p + 2p
78p = 50p + 20p + 5p + 2p + 1p
81p = 50p + 20p + 10p + 1p
38p = 20p + 10p + 5p + 2p + 1p
49p = 20p + 20p + 5p + 2p + 2p

Photocopy Masters

page 10
The next ten

1. $24 + 6 = 30$	2. $15 + 5 = 20$	3. $39 + 1 = 40$
4. $62 + 8 = 70$	5. $58 + 2 = 60$	6. $43 + 7 = 50$
7. $34 + 6 = 40$	8. $16 + 4 = 20$	9. $81 + 9 = 90$
10. $51 + 9 = 60$		

page 11
Adding

1. $16 + 4 = 20$	2. $24 + 6 = 30$	3. $13 + 5 = 18$
4. $16 + 3 = 19$	5. $20 + 4 = 24$	6. $23 + 5 = 28$
7. $24 + 4 = 28$	8. $23 + 3 = 26$	9. $31 + 5 = 36$
10. $40 + 1 = 41$	11. $20 + 9 = 29$	12. $22 + 5 = 27$
13. $23 + 4 = 27$	14. $31 + 7 = 38$	15. $23 + 6 = 29$
16. $22 + 7 = 29$	17. $21 + 9 = 30$	18. $33 + 5 = 38$
19. $31 + 9 = 40$	20. $33 + 4 = 37$	21. $14 + 6 = 20$
22. $21 + 3 = 24$	23. $33 + 6 = 39$	24. $33 + 7 = 40$
25. $25 + 5 = 30$	26. $21 + 6 = 27$	27. $31 + 8 = 39$
28. $24 + 5 = 29$	29. $11 + 7 = 18$	30. $25 + 4 = 29$

page 12
Problem page

1. $13 + 3 = 16$p	2. $23 + 6 = 29$p	3. $13 + 5 = 18$p
4. $26 - 22 = 4$ points	5. $13 + 5 = 18$ years	6. $28 - 6 = 22$ years
7. $10{:}20 + 5\text{min} = 10{:}25$	8. $17 - 6 = 11$	

page 13
Subtracting

1. $25 - 5 = 20$	2. $36 - 6 = 30$	3. $18 - 8 = 10$
4. $47 - 7 = 40$	5. $29 - 9 = 20$	6. $54 - 4 = 50$
7. $26 - 6 = 20$	8. $38 - 8 = 30$	9. $19 - 9 = 10$
10 $42 - 2 = 40$	11. $55 - 5 = 50$	12. $34 - 4 = 30$
13. $36 - 6 = 30$	14. $75 - 5 = 70$	15. $57 - 7 = 50$
16. $24 - 4 = 20$	17. $43 - 3 = 40$	18. $98 - 8 = 90$

Photocopy Masters

page 14
Subtracting

1. $18 - 6 = 12$
2. $15 - 5 = 10$
3. $26 - 4 = 22$
4. $18 - 4 = 14$
5. $29 - 8 = 21$
6. $29 - 6 = 23$
7. $15 - 3 = 12$
8. $25 - 10 = 15$
9. $16 - 5 = 11$
10. $23 - 3 = 20$
11. $27 - 4 = 23$
12. $18 - 7 = 11$
13. $28 - 8 = 20$
14. $29 - 9 = 20$
15. $17 - 6 = 11$
16. $34 - 4 = 30$
17. $35 - 5 = 30$
18. $39 - 8 = 31$
19. $27 - 3 = 24$
20. $26 - 5 = 21$

page 15
Problems page

1. $26 - 4 = 22$ points
2. $35 - 2 = 33$ people
3. $18 - 6 = 12p$
4. $29 - 22 = 7p$
5. $27 - 5 = 22p$
6. $38 - 6 = 32$ years
7. $17 - 4 = 13$ $13 + 17 = 30p$
8. $22 + 4 = 26p$
9. $37 - 5 = 32cm$
10. $19 - 7 = 12$ miles

page 16
Counting in ones

Consecutive numbers 686–728

◉ Number should be alternate colours.

page 17
Counting in tens

1. 16	26	36	46	56	66	76	86	96	106	116
2. 27	37	47	57	67	77	87	97	107	117	
3. 45	55	65	75	85	95	105	115	125	135	
4. 12	22	32	42	52	62	72	82	92	102	112
5. 53	63	73	83	93	103	113	123	143	153	
6. 104	114	124	134	144	154	164	174	184	194	204
7. 87	97	107	117	127	137	147	157	167	177	

Photocopy Masters

page 18
Counting back in tens

1. 63	53	43	33	23	13	
2. 51	41	31	21	11	1	
3. 49	39	29	19	9		
4. 99	89	79	69	59	49	39
5. 175	165	155	145	135	125	115
6. 187	177	167	157	147	137	127
7. 168	158	148	138	128	118	

---

page 19
Ten more, ten less

1. 342	352	362	**2.** 644	654	664
3. 448	458	468	**4.** 490	500	510
5. 822	832	842	**6.** 177	187	197
7. 592	602	612	**8.** 220	230	240
9. 290	300	310	**10.** 719	729	739
11. 517	527	537	**12.** 530	540	550
13. 319	329	339	**14.** 861	871	881
15. 289	299	309			

---

page 20
100 more/100 less

1. 327	427	
2. 416	516	
3. 158	58	
4. 273	373	
5. 495	395	
6. 627	527	
7. 235	335	
8. 348	248	

92	176	255	519	634	328
192	276	355	619	734	428
292	376	455	719	834	528

Photocopy Masters

page 21
Twos

1. 6	8	10	12	14	16	18	20	22	24	26		
2. 14	16	18	20	22	24	26	28	30	32	34		
3. 20	22	24	26	28	30	32	34	36	38	40		
4. 2	4	6	8	10	12	14	16	18	20	22	24	26
5. 10	12	14	16	18	20	22	24	26	28	30	32	
6. 12	14	16	18	20	22	24	26	28	30	32		
7. 16	18	20	22	24	26	28	30	32	34	36		

---

page 22
Twos

1. $4 \times 2 = 8$ **2.** $8 \times 2 = 16$ **3.** $9 \times 2 = 18$ **4.** $10 \times 2 = 20$

5. $2 \times 2 = 4$ **6.** $5 \times 2 = 10$ **7.** $3 \times 2 = 6$ **8.** $1 \times 2 = 2$

9. $6 \times 2 = 12$ **10.** $7 \times 2 = 14$ **11.** $11 \times 2 = 22$ **12.** $12 \times 2 = 24$

13. $50 \times 2 = 100$

---

page 23
Multiplying and dividing

1. $3 \times 6 = 18$ $18 \div 3 = 6$ **2.** $4 \times 4 = 16$ $16 \div 4 = 4$

3. $2 \times 8 = 16$ $16 \div 2 = 8$ **4.** $4 \times 7 = 28$ $28 \div 4 = 7$

5. $5 \times 3 = 15$ $15 \div 5 = 3$ **6.** $3 \times 7 = 21$ $21 \div 3 = 7$

7. $3 \times 5 = 15$ $15 \div 3 = 5$

---

page 24
Multiplying and dividing

1. $3 \times 5 = 15$ $15 \div 5 = 3$ **2.** $4 \times 3 = 12$ $12 \div 3 = 4$

3. $2 \times 4 = 8$ $8 \div 4 = 2$ **4.** $5 \times 4 = 20$ $20 \div 4 = 5$

5. $4 \times 5 = 20$ $20 \div 5 = 4$ **6.** $7 \times 3 = 21$ $21 \div 3 = 7$

7. $8 \times 2 = 16$ $16 \div 2 = 8$ **8.** $1 \times 3 = 3$ $3 \div 3 = 1$

9. $6 \times 10 = 60$ $60 \div 10 = 6$ **10.** $9 \times 2 = 18$ $18 \div 2 = 9$

---

page 25

Doubling

$13 \rightarrow 26$ $11 \rightarrow 22$ $6 \rightarrow 12$ $21 \rightarrow 42$ $8 \rightarrow 16$ $15 \rightarrow 30$ $43 \rightarrow 86$

Photocopy Masters

page 26
Doubling and halving

$9 \rightarrow 18$ $5 \rightarrow 10$ $8 \rightarrow 16$ $10 \rightarrow 20$ $6 \rightarrow 12$ $12 \rightarrow 24$ $11 \rightarrow 22$
$14 \rightarrow 28$ $4 \rightarrow 8$

---

page 27
Fractions **N12**
Halves, quarters and eighths

1–6. 4 parts red, 2 parts blue, 1 part yellow
7–8. 8 parts red, 4 parts blue, 2 parts yellow

---

page 28
Fractions **N12**
Fractions

2. $\frac{1}{3}$ of $9 = 3$ **3.** $\frac{1}{4}$ of $8 = 2$ **4.** $\frac{1}{2}$ of $6 = 3$ **5.** $\frac{1}{3}$ of $12 = 4$

6. $\frac{1}{3}$ of $6 = 2$ **7.** $\frac{1}{7}$ of $12 = 6$ **8.** $\frac{1}{4}$ of $12 = 3$ **9.** $\frac{1}{2}$ of $4 = 2$

10. $\frac{1}{4}$ of $16 = 4$

---

page 29
Fractions **N12**
Fractions

1. $\frac{1}{2}$ of $6 = 3$ **2.** $\frac{1}{3}$ of $9 = 3$ **3.** $\frac{1}{4}$ of $12 = 3$ **4.** $\frac{1}{2}$ of $20 = 10$

5. $\frac{1}{2}$ of $50 = 25$ **6.** $\frac{1}{4}$ of $24 = 6$ **7.** $\frac{1}{4}$ of $20 = 5$ **8.** $\frac{1}{3}$ of $21 = 7$

9. $\frac{1}{2}$ of $30 = 15$ **10.** $\frac{1}{3}$ of $27 = 9$ **11.** $\frac{1}{3}$ of $60 = 20$ **12.** $\frac{1}{4}$ of $100 = 25$

13. $\frac{1}{3}$ of $30 = 10$ **14.** $\frac{1}{2}$ of $14 = 7$ **15.** $\frac{1}{4}$ of $32 = 8$ **16.** $\frac{1}{2}$ of $18 = 9$

---

page 30
Addition/subtraction **N13**
Adding multiples of 10

	top row	second row	bottom row
1.	80		
2.	90		
3.	80		
4.	90		
5.	110	50, 60	
6.	150	70, 80	
7.	150	70, 80	
8.	130	60, 70	

Photocopy Masters

page 30 cont ...

9. 160
10. 170   80, 90   30, 50, 40
11. 150   70, 80   40, 30, 50
12. 170   90, 80   40, 50, 30
13. 150   80, 70   50, 30, 40
14. 160   90, 70   50, 40, 30

page 31
Adding

+	5	10	15	20
5	10	15	20	25
10	15	20	25	30
15	20	25	30	35
20	25	30	35	40

+	10	20	30	40
25	35	45	55	65
35	45	55	65	75
45	55	65	75	85
55	65	75	85	95

+	35	45	55	65
5	40	50	60	70
10	45	55	65	75
15	50	60	70	80
20	55	65	75	85

+	15	25	35	45
25	40	50	60	70
35	50	60	70	80
45	60	70	80	90
55	70	80	90	100

page 32
Subtracting multiples of 10

Answers will vary.

page 33
Adding to 100

1. 26 + 74   2. 37 + 63   3. 44 + 56   4. 19 + 81
5. 52 + 48   6. 96 + 14   7. 32 + 68   8. 71 + 29

Photocopy Masters

page 34
Adding to 100

1. 90 + 10	2. 80 + 20	3. 50 + 50	4. 40 + 60
5. 95 + 5	6. 20 + 80	7. 35 + 65	8. 55 + 45
9. 25 + 75	10. 15 + 85	11. 40 + 60	12. 70 + 30
13. 85 + 15	14. 45 + 55	15. 35 + 65	16. 70 + 30
17. 25 + 75	18. 95 + 5		

---

page 35
Numbers to 1000

1st row	203, 206, 214, 217, 225, 228
2nd row	425, 426, 435, 439, 443, 447
3rd row	616, 620, 623, 628, 636, 639

---

page 36
More and less

1. 229	230	231		2. 145	146	147	
3. 758	759	760		4. 799	800	801	
5. 324	325	326		6. 163	164	165	
7. 998	999	1000		8. 143	243	343	
9. 86	186	286		10. 802	902	1002	
11. 774	874	974		12. 661	761	861	
13. 455	555	655		14. 338	438	538	

---

page 37
3-digit numbers

No answers are required.

---

page 38
The next ten

1. 16 + 4 = 20	2. 23 + 7 = 30	3. 34 + 6 = 40	4. 48 + 2 = 50
5. 68 + 2 = 70	6. 55 + 5 = 60	7. 21 + 9 = 30	8. 46 + 4 = 50
9. 76 + 4 = 80	10. 89 + 1 = 90	11. 38 + 2 = 40	12. 36 + 4 = 40
13. 22 + 8 = 30	14. 17 + 3 = 20	15. 91 + 9 = 100	16. 33 + 7 = 40

Photocopy Masters

page 39
Adding machines

in	23	29	38	17	46	19	28	37
out	27	33	42	21	50	23	32	41

in	37	26	18	45	59	26	35	17
out	42	31	23	50	64	31	40	22

in	13	37	24	48	66	34	55	79
out	20	44	31	55	73	41	62	86

page 40
Missing numbers

1. $27 + 4 = 31$
2. $35 + 7 = 42$
3. $16 + 6 = 22$
4. $28 + 4 = 32$
5. $37 + 5 = 42$
6. $46 + 7 = 53$
7. $19 + 4 = 23$
8. $27 + 5 = 32$
9. $3 + 18 = 21$
10. $7 + 25 = 32$
11. $6 + 27 = 33$
12. $5 + 38 = 43$
13. $4 + 18 = 22$
14. $3 + 29 = 32$
15. $5 + 49 = 54$
16. $4 + 37 = 41$

page 41
Differences

1. $20 - 16 = 4$
2. $31 - 23 = 8$
3. $45 - 36 = 9$
4. $66 - 59 = 7$
5. $69 - 61 = 8$
6. $13 - 3 = 10$
7. $106 - 98 = 8$
8. $79 - 72 = 7$

page 42
Differences

No answers are required.

● 45 & 40, 44 & 39, 43 & 38, 42 & 37, 41 & 36, 40 & 35

Photocopy Masters

page 43
Problem page

1. $8 + 3 = 11$ $8 - 3 = 5$
2. $34 - 27 = 7p$
3. $42 - 38 = 4$ sec
4. $23 - 19 = 4p$
5. $27 + 5 = 32$ $32 + 27 = 59$
6. $21 - 16 = 5cm$
7. $25 - 18 = 7p$
8. $105 - 7 = 98cm$
9. $57 + 5 = 62$
10. $5 - 2 = 3$ 25 (also 85)

---

page 44
Adding money

1. $£1.45 + 30p = £1.75$
2. $20p + 20p = 40p$
3. $20p + £1.20 = £1.40$
4. $£1.20 + £1.45 = £2.65$
5. $30p + £1.45 = £1.75$
6. $30p + 20p = 50p$
7. $£1.20 + £1.45 = £2.65$
8. $£1.20 + 30p = £1.50$
9. $£1.45 + 20p = £1.65$
10. $£1.20 + 30p = £1.50$

---

page 45
Subtracting money

	20p off	50p off	30p off
£1.75	£1.55	£1.25	£1.45
£2.65	£2.45	£2.15	£2.35
£3.99	£3.79	£3.49	£3.69
£1.89	£1.69	£1.39	£1.59
£4.55	£4.35	£4.05	£4.25
£2.85	£2.65	£2.35	£2.55
£5.75	£5.55	£5.25	£5.45
85p	65p	35p	55p

---

page 46
Problems page

1. $£1.65 + 30p = £1.95$
2. $£1.85 - 20p = £1.65$
3. $£3.95 - 20p = £3.75$
4. $£4.55 - 40p = £4.15$
5. $£5.60 + 30p = £5.90$
6. $£3.20 + 30p = £3.50$
7. $£4.70 - £4.40 = 30p$
8. $£5.70 - 10p = £5.60$
9. $£6.70 - 40p = £6.30$
10. $£5.10 + 50p = £5.60$

Photocopy Masters

page 47
Adding two 2-digit numbers

1. 43 + 25 = 68	**2.** 31 + 45 = 76	**3.** 26 + 32 = 58
4. 18 + 21 = 39	**5.** 43 + 55 = 98	**6.** 27 + 52 = 79
7. 36 + 26 = 62	**8.** 58 + 25 = 83	**9.** 18 + 34 = 52
10. 27 + 16 = 43	**11.** 73 + 19 = 92	**12.** 45 + 46 = 91
13. 53 + 28 = 81	**14.** 27 + 37 = 64	**15.** 45 + 28 = 73
16. 35 + 45 = 80	**17.** 67 + 23 = 90	**18.** 73 + 18 = 91
19. 24 + 49 = 73	**20.** 57 + 36 = 93	

page 48
Buying

	−21p	−33p	−15p
£2.46	£2.25	£2.13	£2.31
£3.58	£3.37	£3.25	£3.43
£1.97	£1.76	£1.64	£1.82
£4.46	£4.25	£4.13	£4.31
£3.47	£3.26	£3.14	£3.32
£2.75	£2.54	£2.42	£2.60

page 49
Adding two 2-digit numbers

1. 46 + 35 = 81	**2.** 27 + 35 = 62	**3.** 27 + 38 = 65
4. 27 + 46 = 73	**5.** 59 + 18 = 77	**6.** 27 + 18 = 45
7. 59 + 35 = 94	**8.** 18 + 38 = 56	**9.** 59 + 46 = 105
10. 46 + 38 = 84	**11.** 27 + 59 = 86	

page 50
Adding 9, 19, 29

1. 43 + 19 = 62	**2.** 54 + 9 = 63	**3.** 72 + 29 = 101
4. 81 + 19 = 100	**5.** 68 + 29 = 97	**6.** 29 + 19 = 48
7. 37 + 9 = 46	**8.** 46 + 19 = 65	**9.** 28 + 29 = 57
10. 37 + 29 = 66	**11.** 46 + 9 = 55	**12.** 55 + 19 = 74
13. 87 + 19 = 106	**14.** 64 + 19 = 83	**15.** 82 + 29 = 111
16. 75 + 29 = 104	**17.** 63 + 19 = 82	**18.** 52 + 9 = 61
19. 14 + 39 = 53	**20.** 42 + 39 = 81	

Photocopy Masters

page 51
Adding 9, 19, 29

No answers are required.

🖉 19 + 19 + 9 + 9 + 9

Taking away 19

1. 46 – 19 = 27p	**2.** 54 – 19 = 35p	**3.** 35 – 19 = 16p
4. 26 – 19 = 7p	**5.** 41 – 19 = 22p	**6.** 70 – 19 = 51p
7. 64 – 19 = 45p	**8.** 87 – 19 = 68p	**9.** 63 – 19 = 44p
10. 51 – 19 = 32p	**11.** 45 – 19 = 26p	**12.** 22 – 19 = 3p

Multiples

	mult of 10	mult of 50	mult of 100	mult of 2	mult of 5
630	✓	✗	✗	✓	✓
950	✓	✓	✗	✓	✓
35	✗	✗	✗	✗	✓
180	✓	✗	✗	✓	✓
220	✓	✗	✗	✓	✓
95	✗	✗	✗	✗	✓
20	✓	✗	✗	✓	✓
46	✗	✗	✗	✓	✗
17	✗	✗	✗	✗	✗
100	✓	✓	✓	✓	✓

Odd and even

1. 15 + 5 = 20	**2.** 16 + 3 = 19	**3.** 17 + 5 = 22	**4.** 8 + 8 = 16
5. 13 + 7 = 20	**6.** 12 + 3 = 15	**7.** 21 + 3 = 24	**8.** 21 + 2 = 23
9. 16 + 5 = 21	**10.** 17 + 4 = 21	**11.** 11 + 4 = 15	**12.** 12 + 2 = 14
13. 16 + 2 = 18	**14.** 17 + 1 = 18		

Photocopy Masters

page 55
Missing numbers

1. 30 + 70 = 100	**2.** 60 + 40 = 100	**3.** 80 + 20 = 100
4. 10 + 90 = 100	**5.** 50 + 50 = 100	**6.** 40 + 60 = 100
7. 70 + 30 = 100	**8.** 20 + 80 = 100	**9.** 90 + 10 = 100
10. 35 + 65 = 100	**11.** 25 + 75 = 100	**12.** 85 + 15 = 100
13. 45 + 55 = 100	**14.** 65 + 35 = 100	**15.** 15 + 85 = 100
16. 75 + 25 = 100	**17.** 5 + 95 = 100	**18.** 55 + 45 = 100

---

page 56
Adding to 100

11, 89	65, 35	25,75	46, 54	72, 28	31, 69
45, 55	8, 92	10, 90	20, 80		

---

page 57
Adding

1. 69 + 30 = 99	**2.** 61 + 41 = 102	**3.** 78 + 20 = 98
4. 34 + 65 = 99	**5.** 25 + 74 = 99	**6.** 85 + 16 = 101
7. 44 + 54 = 98	**8.** 66 + 35 = 101	**9.** 16 + 85 = 101
10. 76 + 26 = 102	**11.** 96 + 5 = 101	**12.** 56 + 45 = 101
13. 35 + 75 = 110	**14.** 65 + 45 = 110	**15.** 18 + 90 = 108
16. 32 + 73 = 105	**17.** 41 + 62 = 103	**18.** 23 + 81 = 104
19. 72 + 31 = 103	**20.** 82 + 21 = 103	

---

page 58
Doubling machine

in	21	24	32	23	14	41	31	42
out	42	48	64	46	28	82	62	84

in	11	34	22	12	44	33	13	43
out	22	68	44	24	88	66	26	86

in	25	5	45	75	15	65	35	55
out	50	10	90	150	30	130	70	110

Photocopy Masters

page 59
Adding near doubles

1. $25 + 26 = 51$
2. $14 + 15 = 29$
3. $35 + 37 = 72$
4. $11 + 12 = 23$
5. $13 + 14 = 27$
6. $22 + 23 = 45$
7. $33 + 34 = 67$
8. $44 + 45 = 89$
9. $15 + 16 = 31$
10. $23 + 24 = 47$
11. $41 + 42 = 83$
12. $35 + 36 = 71$
13. $12 + 14 = 26$
14. $24 + 26 = 50$
15. $36 + 37 = 73$
16. $45 + 46 = 91$
17. $42 + 44 = 86$
18. $31 + 33 = 64$

page 60
Tens

$3 \rightarrow 30$ $10 \rightarrow 100$ $9 \rightarrow 90$ $5 \rightarrow 50$ $4 \rightarrow 40$
$2 \rightarrow 20$ $6 \rightarrow 60$ $7 \rightarrow 70$ $8 \rightarrow 80$ $1 \rightarrow 10$

page 61
Fives

1. $3 \times 5 = 15$
2. $4 \times 5 = 20$
3. $6 \times 5 = 30$
4. $4 \times 5 = 20$
5. $5 \times 5 = 25$
6. $2 \times 5 = 10$
7. $2 \times 5 = 10$
8. $9 \times 5 = 45$
9. $10 \times 5 = 50$
10. $8 \times 5 = 40$
11. $12 \times 5 = 60$
12. $8 \times 5 = 40$
13. $6 \times 5 = 30$
14. $10 \times 5 = 50$
15. $1 \times 5 = 5$

page 62
Multiplying

1. $3 \times 5 = 15$
2. $3 \times 4 = 12$
3. $5 \times 2 = 10$
4. $3 \times 6 = 18$
5. $5 \times 4 = 20$
6. $6 \times 4 = 24$
7. $2 \times 4 = 8$
8. $4 \times 4 = 16$
9. $4 \times 3 = 12$
10. $6 \times 3 = 18$

page 63
Multiplying

1. A & F $3 \times 4 = 4 \times 3$
2. B & H $6 \times 3 = 3 \times 6$
3. C & I $2 \times 7 = 7 \times 2$
4. D & J $2 \times 3 = 3 \times 2$
5. E & G $4 \times 5 = 5 \times 4$

Photocopy Masters

page 64
Twos and threes

I	2
3	4
5	6
7	8
q	10
II	12
13	14
15	16
17	18
19	20

I	2	3
4	5	6
7	8	q
10	II	12
13	14	15
16	17	18
19	20	21
22	23	24
25	26	27
28	29	30

page 65
Threes

12	24	6
3	15	21
18	27	q

q	21	15
18	24	6
30	12	27

36	24	12
27	q	33
15	30	21

page 66
Fractions

 $\frac{3}{4}$
 $\frac{2}{3}$
 $\frac{7}{8}$
 $\frac{2}{5}$
 $\frac{2}{4}$

 $\frac{4}{6}$
 $\frac{7}{12}$
 $\frac{3}{8}$
 $\frac{5}{8}$
 $\frac{5}{6}$

Photocopy Masters
page 66 cont ...

 $\frac{2}{6}$

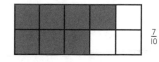 $\frac{7}{10}$

---

page 67
Fractions

1. $\frac{3}{4}$ of 8 = 6   2. $\frac{2}{3}$ of 9 = 6   3. $\frac{7}{8}$ of 24 = 21   4. $\frac{3}{5}$ of 10 = 6

5. $\frac{5}{6}$ of 18 = 15   6. $\frac{1}{4}$ of 16 = 4   7. $\frac{2}{5}$ of 15 = 6   8. $\frac{1}{8}$ of 16 = 2

---

page 68
Nearest ten

blue cars	64	63	59	56	61
green cars	31	32	34	28	
orange cars	6	14	7	9	
yellow cars	89	92	91	85	94

---

page 69
Nearest ten

1. 555	556	557	558	559	<u>560</u>	561	562	563	564
2. 475	476	477	478	479	<u>480</u>	481	482	483	484
3. 325	326	327	328	329	<u>330</u>	331	332	333	334
4. 205	206	207	208	209	<u>210</u>	211	212	213	214
5. 795	796	797	798	799	<u>800</u>	801	802	803	804
6. 945	946	947	948	949	<u>950</u>	951	952	953	954
7. 685	686	687	688	689	<u>690</u>	691	692	693	694

---

page 70
Nearest hundred

Any of the following are possible.

1. 294   300   2. 297   300   3. 249   200

4. 247   200   5. 274   300   6. 279   300

Photocopy Masters

page 70 cont ...

7. 427	400	**8.** 429	400	**9.** 472	500
10. 479	500	**11.** 492	500	**12.** 497	500
13. 724	700	**14.** 729	700	**15.** 742	700
16. 749	700	**17.** 792	800	**18.** 794	800
19. 924	900	**20.** 927	900	**21.** 942	900
22. 947	900	**23.** 972	1000	**24.** 974	1000

page 71
Addition tables

Addition/subtraction N30

+	10	60	30	50
50	60	110	80	100
20	30	80	50	70
40	50	100	70	90
30	40	90	60	80

+	15	35	5	25
20	35	55	25	45
30	45	65	35	55
10	25	45	15	35
40	55	75	45	65

+	45	35	5	25
15	60	50	20	40
35	80	70	40	60
25	70	60	30	50
55	100	90	60	80

+	25	20	40	45
45	70	65	85	90
30	55	50	70	75
35	60	55	75	80
50	75	70	90	95

Photocopy Masters

page 72
Problem page

1. 30 + 25 = 55cm
2. 45 – 15 = 30p
3. 50 – 35 = 15 boys
4. 35 + 25 = 60 children
5. 85 – 60 = 25 miles
6. 70 – 45 = 25 stickers
7. 20 + 30 + 15 = 65 points
8. 35 + 25 + 20 = 80   £1 – 80p = 20p

---

page 73
Subtracting 20, 30, 50

11	5	26
9	43	14
32	7	55

9	61	26
52	18	21
34	40	43

13	39	42
22	30	25
7	34	43

---

page 74
Taking away 30

1. 59p – 30p = 29p
2. 75p – 30p = 45p
3. 48p – 30p = 18p
4. 61p – 30p = 31p
5. 56p – 30p = 26p
6. 72p – 30p = 42p
7. 43p – 30p = 13p
8. 67p – 30p = 37p
9. 54p – 30p = 24p

---

page 75
Problem page

1. 45 – 20 = 25p
2. 76 – 40 = 36p
3. 38 – 21 = 17 points
4. 65 – 31 = 34cm
5. 54 – 33 = 21 cards
6. 48 – 13 = 35 children
7. 67 + 22 = 89p
8. 76 – 42 = 34 sheep

---

page 76
Difference tables

d	17	23	19	22
16	1	7	3	6
18	1	5	1	4
25	8	2	6	3
21	4	2	2	1

Photocopy Masters
page 76 cont ...

d	31	34	27	30
28	3	6	I	2
33	2	I	6	3
29	2	5	2	I
32	I	2	5	2

d	42	39	44	37
36	6	3	8	I
43	I	4	I	6
38	4	I	6	I
46	4	7	2	q

d	62	58	65	55
59	3	I	6	4
63	I	5	2	8
57	5	I	8	2
64	2	6	I	q

---

page 77
Problem page

I. 13 − 9 = 4 years
2. 23 − 17 = 6 children
3. 32 − 29 = 3p
4. 43 − 5 = 38cm
5. 26 − 19 = 7cm
6. 54 − 48 = 6 paces
7. 73 − 65 = £8
8. 23 − 4 = 19

---

page 78
Adding three 2-digit numbers

No answers are required.

---

page 79
Adding

I. 35 + 15 + 7 = 57
2. 25 + 25 + 12 = 62
3. 45 + 25 + 14 = 84
4. 35 + 35 + 13 = 83
5. 5 + 45 + 16 = 66
6. 35 + 5 + 17 = 57
7. 65 + 15 + 13 = 93
8. 25 + 75 + 15 = 115
9. 35 + 5 + 21 = 61
10. 25 + 15 + 22 = 62
II. 45 + 35 + 23 = 103
12. 75 + 15 + 11 = 101
13. 25 + 35 + 32 = 92
14. 65 + 25 + 14 = 104
15. 35 + 45 + 15 = 95
16. 85 + 5 + 16 = 106

Photocopy Masters

page 80
Adding three 2-digit numbers
1. $25 + 31 + 12 = 68$
2. $12 + 29 + 46 = 87$
3. $12 + 37 + 44 = 93$
4. $31 + 37 + 27 = 95$
5. $25 + 37 + 46 = 108$
6. $44 + 27 + 46 = 117$
7. $58 + 37 + 29 = 124$
8. $25 + 58 + 44 = 127$

page 81
Missing numbers
1. $135 + 5 = 140$
2. $127 + 3 = 130$
3. $156 + 4 = 160$
4. $238 + 2 = 240$
5. $327 + 3 = 330$
6. $514 + 6 = 520$
7. $626 + 4 = 630$
8. $329 + 1 = 330$
9. $252 + 8 = 260$
10. $148 + 2 = 150$
11. $523 + 7 = 530$
12. $485 + 5 = 490$
13. $277 + 3 = 280$
14. $191 + 9 = 200$
15. $138 + 2 = 140$
16. $274 + 6 = 280$

page 82
Adding
1. $126 + 13 = 139$
2. $361 + 14 = 375$
3. $235 + 21 = 256$
4. $182 + 16 = 198$
5. $424 + 31 = 455$
6. $423 + 22 = 445$
7. $313 + 43 = 356$
8. $215 + 51 = 266$
9. $506 + 42 = 548$
10. $374 + 13 = 387$
11. $532 + 46 = 578$
12. $651 + 24 = 675$
13. $443 + 25 = 468$
14. $264 + 14 = 278$
15. $125 + 64 = 189$
16. $376 + 22 = 398$

page 83
Problem page
1. $135 + 20 = 155$
2. $216 + 7 = 223$ marbles
3. $325 + 8 = 333$ days
4. $135 + 14 = 149$ points
5. $162 - 156 = 6p$
6. $128 + 30 = 158$ children
7. $458 - 432 = 26p$
8. $579 - 547 = 32$ stamps

page 84
Adding multiples of 100
Answers will vary.

Photocopy Masters

page 85
Adding 3-digit numbers

Answers will vary.

---

page 86
Subtracting 3-digit numbers

1. $354 - 110 = 244$
2. $426 - 220 = 206$
3. $473 - 150 = 323$
4. $158 - 140 = 18$
5. $274 - 130 = 144$
6. $229 - 110 = 119$
7. $634 - 220 = 414$
8. $748 - 330 = 418$
9. $537 - 410 = 127$
10. $486 - 220 = 266$
11. $194 - 150 = 44$
12. $288 - 160 = 128$
13. $365 - 230 = 135$
14. $644 - 420 = 224$

---

page 87
Odd and even

Completed 1–100 grid with alternate columns of red and blue.

---

page 88
Money

1. £5 = 500p
2. £3 = 300p
3. £1.50 = 150p
4. £2.50 = 250p
5. £4 = 400p
6. £6.50 = 650p
7. £7.25 = 725p
8. £8.10 = 810p
9. £9.35 = 935p
10. £1.17 = 117p
11. £3.45 = 345p
12. £7.03 = 703p

---

page 89
Tens

1. $9 \times 10 = 90$
2. $7 \times 10 = 70$
3. $3 \times 10 = 30$
4. $18 \times 10 = 180$
5. $16 \times 10 = 160$
6. $12 \times 10 = 120$
7. $240 \div 10 = 24$
8. $360 \div 10 = 36$
9. $910 \div 10 = 91$
10. $56 \times 10 = 560$
11. $8 \times 10 = 80$
12. $2 \times 10 = 20$
13. $430 \div 10 = 43$
14. $290 \div 10 = 29$
15. $31 \times 10 = 310$
16. $100 \div 10 = 10$
17. $15 \times 10 = 150$
18. $380 \div 10 = 38$

Photocopy Masters

page 90
Multiplying

1. $2 \times 30 = 60$	**2.** $3 \times 20 = 60$	**3.** $4 \times 20 = 80$
4. $2 \times 40 = 80$	**5.** $5 \times 40 = 200$	**6.** $3 \times 40 = 120$
7. $6 \times 30 = 180$	**8.** $4 \times 30 = 120$	**9.** $7 \times 30 = 210$
10. $8 \times 20 = 160$	**11.** $2 \times 50 = 100$	**12.** $3 \times 60 = 180$
13. $5 \times 30 = 150$	**14.** $6 \times 20 = 120$	**15.** $9 \times 20 = 180$
16. $4 \times 40 = 160$	**17.** $7 \times 40 = 280$	**18.** $5 \times 70 = 350$
19. $3 \times 50 = 150$	**20.** $4 \times 60 = 240$	

page 91
Multiplying

Answers will vary.

page 92
Threes and fours

No answers are required.

● Threes: 3, 6, 9, 12, 15, 18, 21, 24, 27, 30, 33, 36, 39, 42, 45, 48
Fours: 4, 8, 12, 16, 20, 24, 28, 32, 36, 40, 44, 48
(12, 24, 36 and 48 will be coloured in both colours)

page 93
Fours

16	8	24
12	32	4
28	20	36

8	36	16
32	20	40
24	12	28

24	40	12
16	32	28
36	44	20

page 94
Dividing

1. $10 \div 2 = 5$	**2.** $8 \div 2 = 4$	**3.** $12 \div 3 = 4$	**4.** $9 \div 3 = 3$
5. $6 \div 3 = 2$	**6.** $12 \div 4 = 3$	**7.** $18 \div 2 = 9$	**8.** $27 \div 3 = 9$
9. $21 \div 3 = 7$			

Money

1. £5 = 500p	**2.** £3 = 300p	**3.** £1.50 = 150p
4. £2.50 = 250p	**5.** £4 = 400p	**6.** £6.50 = 650p
7. £7.25 = 725p	**8.** £8.10 = 810p	**9.** £9.35 = 935p
10. £1.17 = 117p	**11.** £3.45 = 345p	**12.** £7.03 = 703p

Photocopy Masters

page 95
Remainders

1. $9 \div 4 = 2r1$	2. $7 \div 2 = 3r1$	3. $4 \div 3 = 1r1$	4. $11 \div 4 = 2r3$
5. $11 \div 3 = 3r2$	6. $19 \div 2 = 9r1$	7. $17 \div 5 = 3r2$	8. $47 \div 10 = 4r7$
9. $8 \div 3 = 2r2$	10. $17 \div 4 = 4r1$	11. $13 \div 2 = 6r1$	12. $38 \div 5 = 7r3$
13. $33 \div 10 = 3r3$	14. $26 \div 4 = 6r2$	15. $17 \div 3 = 5r2$	16. $46 \div 5 = 9r1$

page 96
Fractions

1. $\frac{4}{8} = \frac{1}{2}$	2. $\frac{6}{8} = \frac{3}{4}$	3. $\frac{5}{8}$	4. $\frac{3}{8}$	5. $\frac{2}{8} = \frac{1}{4}$
6. $\frac{4}{8} = \frac{1}{2}$	7. $\frac{4}{8} = \frac{1}{2}$	8. $\frac{2}{8} = \frac{1}{4}$	9. $\frac{4}{8} = \frac{1}{2}$	10. $\frac{4}{8} = \frac{1}{2}$

page 97
Fractions

1. $\frac{1}{2} = \frac{3}{6}$	2. $\frac{1}{3} = \frac{2}{6}$	3. $\frac{3}{4} = \frac{6}{8}$	4. $\frac{2}{3} = \frac{4}{6}$	5. $\frac{1}{4} = \frac{2}{8}$
6. $\frac{1}{2} = \frac{4}{8}$	7. $\frac{2}{5} = \frac{4}{10}$			

page 98
Fractions

1. $\frac{1}{3} = \frac{2}{6}$	2. $\frac{2}{4} = \frac{3}{6}$	3. $\frac{2}{3} = \frac{4}{6}$	4. $\frac{3}{6} = \frac{6}{12}$	5. $\frac{3}{4} = \frac{9}{12}$
6. $\frac{1}{2} = \frac{2}{4}$	7. $\frac{1}{2} = \frac{3}{6}$	8. $\frac{5}{6} = \frac{10}{12}$	9. $\frac{1}{4} = \frac{3}{12}$	10. $\frac{1}{3} = \frac{4}{12}$
11. $\frac{1}{2} = \frac{6}{12}$	12 $1 = \frac{6}{6}$			

page 99
Adding

1. $335 + 107 = 442$	2. $534 + 104 = 638$	3. $426 + 104 = 530$
4. $625 + 205 = 830$	5. $335 + 205 = 540$	6. $303 + 426 = 729$
7. $625 + 104 = 729$	8. $205 + 426 = 631$	9. $534 + 303 = 837$
10. $107 + 625 = 732$	11. $335 + 303 = 638$	12. $534 + 205 = 739$

page 100
Adding

Answers will vary.

Photocopy Masters

page 101
Adding three 3-digit numbers

I. 707	**2.** 675	**3.** 683	**4.** 691	**5.** 682
6. 701	**7.** 778	**8.** 804	**9.** 783	**10.** 819
II. 827	**12.** 795			

page 102
Adding three 2-digit numbers

No answers are required.

page 103
Centimetres

Answers will vary.

page 104
Metres

Length

I. 1m	**2.** 2m	**3.** any possible
4. 3m	**5.** 3m	**6.** any possible >6m
7. any possible	**8.** any possible	

page 105
Metres and centimetres

Length

I. 1m 35cm = 135cm	**2.** 2m 10cm = 210 cm	**3.** 3m = 300 cm
4. $1\frac{1}{2}$ m = 150 cm	**5.** $2\frac{1}{4}$ m = 225 cm	**6.** 4m 58cm = 458 cm
7. 6m 7cm = 607 cm	**8.** 9m 20cm = 920 cm	**9.** 245cm = 2m 45 cm
10. 325cm = 3m 25 cm	**II.** 400cm = 4m	**12.** 150cm = 1m 50 cm
13. 208cm = 2m 8 cm	**14.** 306cm = 3m 6 cm	**15.** 65cm = 0m 65 cm
16 $2\frac{1}{2}$ m = 2m 50 cm		

page 106
Quarter past, half past, quarter to

Time

I. 2:15	**2.** 11:15	**3.** 1:45	**4.** 8:00	**5.** 3:45
6. 11:30				

Photocopy Masters

page 106 cont ...

7.

8.

9.

10.

11.

12.

---

page 107
Timing

1. 20 minutes	**2.** 10 minutes	**3.** 15 minutes	**4.** 5 minutes
5. 35 minutes	**6.** 60 minutes	**7.** 30 minutes	**8.** 25 minutes
9. 50 minutes	**10.** 55 minutes	**11.** 45 minutes	**12.** 40 minutes

---

page 108
5 minutes

1. 8:00	**2.** 8:05	**3.** 8:10	**4.** 8:15	**5.** 8:20
6. 8:25	**7.** 8:30	**8.** 8:35	**9.** 8:40	**10.** 8:45
11. 8:50	**12.** 8:55			

---

page 109
5 minutes

1. twenty past 3   **2.** twenty to 6   **3.** five past 7

Photocopy Masters
page 109 cont ...

4. twenty-five to 2

5. ten past 6

6. ten to 3

7. twenty-five past 9

8. five to 5

9. quarter to 9

page 110
Litres and millilitres

I. 1 l 200ml = 1200ml
2. 2 l 400 ml = 2400ml
3. 5 l = 5000ml
4. $3\frac{1}{2}$ l = 3500ml
5. 1 l 700ml = 1700ml
6. 4 l 600ml = 4600ml
7. $2\frac{1}{2}$ l = 2500ml
8. $1\frac{1}{2}$ l = 1500ml
9. 1500ml = 1 l 500ml
10. 2300ml = 2l 300ml
II. 700ml = 0 l 700ml
12. 3000ml = 3 l
13. $1\frac{1}{2}$ l = 1l 500ml
14. 5600ml = 5 l 600ml
15. 1100ml = 1l 100ml
16. $4\frac{1}{2}$ l = 4l 500ml

page 111
Hours and minutes

I. 1hr 10min = 70 min
2. 2hr 30 min = 150 min
3. 1hr 55 min = 115 min
4. 2hr 5 min = 125 min
5. $1\frac{1}{2}$ hr = 90 min
6. $2\frac{1}{4}$ hr = 135 min
7. $3\frac{1}{2}$ hr = 210 min
8. 4hr = 240 min
9. 90 min = 1hr 30 min
10. 130 min = 2hr 10 min
II. 65 min = 1hr 5 min
12. 200 min = 3hr 20 min
12. 100 min = 1hr 40 min
14. 50 min = 0hr 50 min
15. $1\frac{1}{2}$ hr = 1hr 30 min
16. $3\frac{1}{4}$ hr = 3 hr 15 min

Photocopy Masters

page 112
Grams (g)

Answers will vary.

---

page 113
Grams (g) and kilograms (kg)

I. Ig **2.** 120g **3.** 500g **4.** 1500kg **5.** Ikg

500 + 100 + 50 = 650g
200 + 50 + 20 = 270g
500 + 200 + 100 + 50 = 850g
Ikg + 500 + 20 = 1520g
Ikg + 500 + 200 + 100 + 50 + 20 = 1870g

---

page 114
Months

I. January **2.** December **3.** April **4.** October
5. April **6.** August **7.** February **8.** October
9. April **10.** November **II.** June, July, Aug
12. Dec, Jan, Feb **13.** various

---

page 115
2-d shape

I.

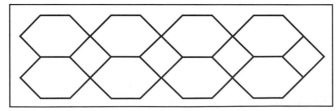

2.

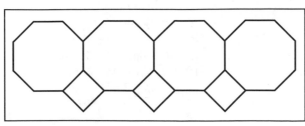

Photocopy Masters

S1

page 115 cont ...

3.

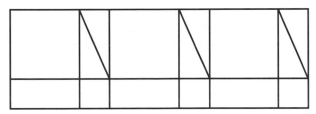

---

page 116
2-d shape

2-d shape S1

I. rectangle	**2.** triangle	**3.** octagon	**4.** pentagon
5. square	**6.** hexagon	**7.** triangle	**8.** pentagon
q. hexagon			

---

page 117
Symmetry

Symmetry S2

I.

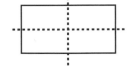

2.

3.

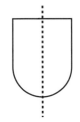

4.

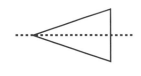

5.

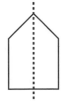

6.

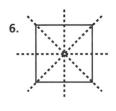

Shapes with more than I line of symmetry: I, 2, 6

Photocopy Masters

page 118
Symmetry

1.

2.

3.

4.

5.

6.

7.

---

page 119

North, east, south, west

1. east 2 spaces   south 2 spaces   west 3 spaces   south 2 spaces
2. east 4 spaces   north 2 spaces   east 3 spaces   south 6 spaces
3. north 4 spaces   west 4 spaces   south 5 spaces   east 6 spaces
   north 3 spaces
4. east 3 spaces   north 3 spaces   east 2 spaces   south 2 spaces
   east 2 spaces   north 4 spaces
5. west 3 spaces   north 2 spaces   west 1 space   south 4 spaces
   east 6 spaces   north 3 spaces

Photocopy Masters

page 119 cont ...

reverses

1. north 2 spaces   east 3 spaces   north 2 spaces   west 2 spaces
2. north 6 spaces   west 3 spaces   south 2 spaces   west 4 spaces
3. south 3 spaces   west 6 spaces   north 5 spaces . east 4 spaces
      south 4 spaces
4. south 4 spaces   west 2 spaces   north 2 spaces   west 2 spaces
      south 3 spaces   west 3 spaces
5. south 3 spaces   west 6 spaces   north 4 spaces   east 1 space
      south 2 spaces   east 3 spaces

---

page 120
North, east, south, west

1. south	2. west	3. east	4. north	5. east	6. south
7. east	8. south	9. west	10. north	11. west	12. north

---

page 121
Right-angles

coloured blue   3,
coloured yellow   1, 5, 6, 7, 8, 9
coloured green   2, 4,

---

page 122
Turning north, east, south, west

No answers are required.

---

page 123
Position

1–8.

9. Joe
10. Ann
11. Lisa

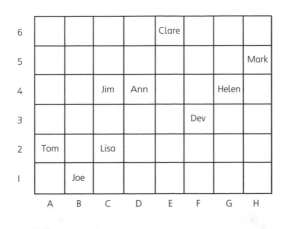

Photocopy Masters

page 124
Tally charts

Fruit	Tallies	Total
apple	卌	5
orange	卌 IIII	9
banana	卌 II	7
plum	卌 I	6
pineapple	IIII	4

1. orange   2. pineapple

page 125
Frequency table

Frequency tables D2

Vowel	Frequency
a	13
e	18
i	6
o	14
u	8

1. e   2. i

page 126
Bar graph

Bar graphs D3

1. 24 books   2. 28 books   3. Wednesday   4. Tuesday
5. Thursday   6. Friday   7. 2 days   8. 2 days
9. Saturday   10. 164 books

page 127
Pictograph

Pictographs D4

1. Autumn   2. Spring   3. Summer   4. Winter
5. Autumn   6. Spring   7. Summer   8. Spring
9. Spring   10. Autumn

24 children altogether

Contents